NONLINEAR OPTICS FOR THE INFORMATION SOCIETY

Nonlinear Optics for the Information Society

Proceedings of the Third Annual Meeting of the COST Action P2, held in Enschede, The Netherlands, 26–27 October 2000

Edited by

A. DRIESSEN
MESA+ Research Institute,
University of Twente,
Enschede, The Netherlands

KLUWER ACADEMIC PUBLISHERS
DORDRECHT / BOSTON / LONDON

A C.I.P. Catalogue record for this book is available from the Library of Congress.

ISBN 978-90-481-5876-8 ISBN 978-94-015-1267-1 (eBook)

DOI 10.1007/978-94-015-1267-1

Published by Kluwer Academic Publishers,
P.O. Box 17, 3300 AA Dordrecht, The Netherlands.

Sold and distributed in North, Central and South America
by Kluwer Academic Publishers,
101 Philip Drive, Norwell, MA 02061, U.S.A.

In all other countries, sold and distributed
by Kluwer Academic Publishers,
P.O. Box 322, 3300 AH Dordrecht, The Netherlands.

Printed on acid-free paper

Preface

MESA$^+$ is the largest research institute of the University of Twente. This multidisciplinary institute focuses on the materials, technologies and systems of the information society. Micro System technology, Micro-Electronics and Materials Science are key words. The institute combines equal research contributions from the disciplines of Electrical Engineering, Applied Physics and Chemical Technology.

MESA$^+$ has a strong international reputation, based on its ongoing strength in research which has its roots in the 70's. MESA$^+$ was formed in 1999, as a merger of MESA and the Center for Materials Research (CMO). The institute has a staff of about 380 people, of which approximately 280 are scientists. In full-time equivalents, the total number of employees is around 300.

The present volume of the MESA$^+$ monographs contains the Proceedings of Nonlinear Optics for the Information Society, NOIS 2000, the third annual workshop of the COST Action P2, held 26 - 27 October 2000 at the University of Twente, The Netherlands. COST is a framework for scientific and technical cooperation, allowing the co-ordination of national research on a European level. COST Actions consist of basic and pre-competitive research as well as activities of more direct public utility. The goal of COST is to ensure that Europe holds a strong position in the field of scientific and technical research for peaceful purposes, by increasing European co-operation and interaction in this field

NOIS 2000 gives an overview of work carried out in the COST P2 Action in physics dealing with Applications of Nonlinear Optical Phenomena. Papers were presented in the following fields:

- nonlinear optical effects, devices, materials
- optical memories
- ultra-fast electro-optic or all-optical data processing
- nonlinear optics for quantum communication
- nonlinear effects in semiconductor lasers

The organizers wish to thank the COST Secretariat, the Dutch Foundation FOM and the University of Twente for financial support of the meeting, and Kluwer Academic Publishers for showing continuing interest. The help of Hugo J.W.M. Hoekstra and Arne Leinse in the completion of this edition is gratefully acknowledged.

We are pleased to present you the Proceedings of NOIS 2000 in this MESA$^+$ monograph.

Enschede, May 2001

David N. Reinhoudt
Scientific Director of MESA$^+$

Alfred Driessen
Editor

CONTENTS

NOVEL APPLICATIONS OF NONLINEAR MATERIALŞ/PBG

OPTICAL MEMORIES

POSTER SESSIONS

THE ROLE OF NONLINEAR OPTICAL DEVICES IN THE OPTICAL COMMUNICATION AGE

L. DALTON
Department of Chemistry, University of Washington
Box 351700, Bagley Hall 202D, Seattle, WA 98195-1700, USA

Abstract or full article not available

http://depts.washington.edu/chemfac/dalton.html

A. Driessen (ed.), *Nonlinear Optics for the Information Society*, 1.

NON-LINEAR OPTICAL TECHNOLOGY IN PRESENT AND FUTURE TRANSMISSION SYSTEMS AND OPTICAL NETWORKS

D. ERASME
Ecole Nationale Superieure des Telecommunications CNRS URA 0820, Département Communications et Electronique, 46, rue Barrault 75634 Paris CEDEX 13, France

Abstract

Within the spectacular and notorious explosion of the information technology economy, the part dedicated to hardware developments remains partially hidden. Nevertheless, the photonic technology industry strongly participates to the present expansion especially in the telecommunication domain. The maturing of WDM technologies has provided the key towards success. The requirements for data transmission capabilities related to the development of the Internet and other network exchanges has been fulfilled. A major characteristic related to the development relies in the long-term research and component and device design required for the new generation of systems. Although original applications of optics was concentrated on point-to-point transmission, advanced optical network architectures foresee a wide use of the "optical parameter" as a main feature for the required exchanges between the various network levels (local, metropolitan, backbone...). It is clear that photonic components are the fundamental building block that will have to perform the functions now carried out in the electrical domain, in order to fully explore and implement the all-optical networks. The strategic choice of transmitted signal granularity in every network node determines the optical/electronic and the WDM/OTDM alternatives and relies on the new component performances. A first generation of devices for optical routing and processing has been proposed. It is mainly based on semiconductor optical amplifiers (SOAs) that provide the strong non-linear operation required through cross-phase and cross-gain modulation and four-wave-mixing. Non-linearities in optical fibers (self-phase modulation, Raman amplification...) have also lead to some interesting proposals. Major achievements have been realized in the field of wavelength converters and 3R regenerators. These devices and sub-systems are ready to be used in industrial deployments. Present and future research should focus on the enhancement of non-linear operation (through strong light spatial confinement for example...) or of device speed, while keeping in minds the requirements for integration. Photonic band-gap structures in semiconductors and microstructured optical fibers already appears as powerful candidates for a new generation of components. New semiconductor materials and extensive use of polymers enhance the application fields of photonics well beyond the telecom application domain. "The new generation optical communications networks are being built using devices

A. Driessen (ed.), Nonlinear Optics for the Information Society, 3–22.

that emerge from fundamental research from the last ten years… Pursuing this research into basic physics may lead to a completely new area of communication and computing, with perspectives that we are not likely to imagine today" (OPTIMIST)

OPTIMIST

Non-linear optical technology in present and future transmission systems and optical networks.

Didier ERASME
École Nat. Sup. des Télécommunications Paris
Yves Jaouën, Jacques Benoit
Henrik Poulsen, Bjarne Tromborg TUD
Ann Ackaert, P. Lagasse IMEC

http://www.ist-optimist.org/

OPTIMIST

Consortium

- ☺ Interuniversitair Microelectronica Centrum (B)
  - ⇨ TELSCOM CONSULTING GmbH (CH)
- ☺ University of Essex (UK)
- ☺ Research Center COM, Technical University of Denmark (DK)
- ☺ École Nationale Supérieure des Télécommunications(F)
- ☺ Heinrich-Hertz-Institut fur Nachrichtentechnik Berlin GmbH (D)

NOIS COST P2 Workshop — 26 October 2000

External Interactions

OPTIMIST

- ☺ Objective: interact / exchange information with
  - ⇨ COST actions in the photonic area
  - ⇨ EURESCOM projects
  - ⇨ National Initiatives EU Member States
  - ⇨ IETF
  - ⇨ FSAN
  - ⇨ OITDA / OIDA
  - ⇨ others....
- ☺ Action: listing of interesting / possible interactions
  - ⇨ forum for information dissemination
  - ⇨ medium for completing views on the *TTx*'s

NOIS COST P2 Workshop 26 October 2000

Dissemination

OPTIMIST

- ☺ Information dissemination in-/outside IST
- ☺ OPTIMIST Quarterly Newsletter
  - ⇨ electronic distribution of OQN / linked to website
    - *announcements for & minutes of workshops*
    - *information on "cluster" projects*
    - *updated info on "Technology Trends"*
    - *events / "commissions corner"*
    - *links to other projects / organisations*

http://www.ist-optimist.org/

NOIS COST P2 Workshop 26 October 2000

TRANSMISSION

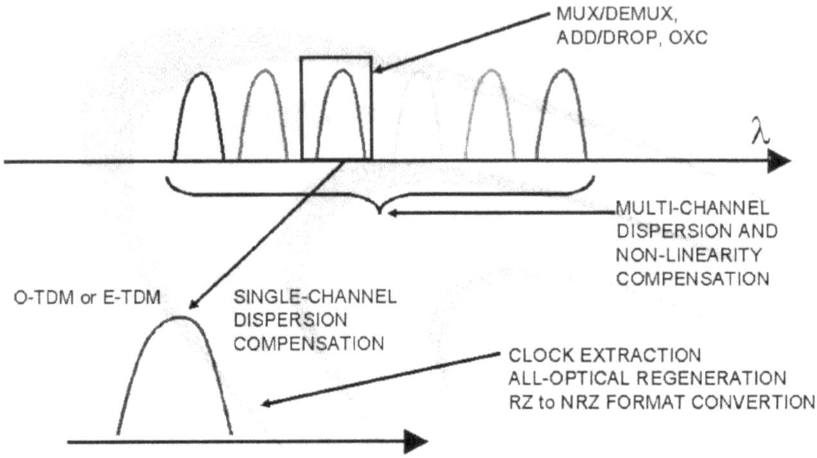

NOIS COST P2 Workshop — 26 October 2000

Optical Bandwidth

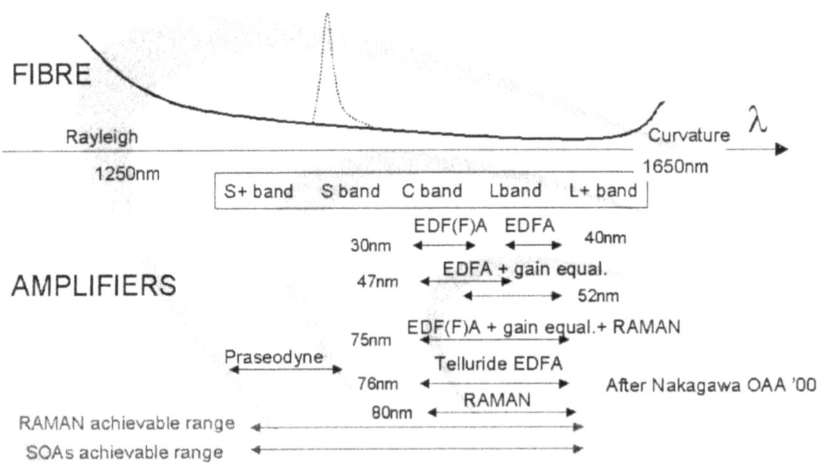

NOIS COST P2 Workshop — 26 October 2000

Outline of the talk

OPTIMIST

☺State-of-the-art in Optical Telecommunications

☺Non-linear optics in today 's systems

☺Perspective for non-linear optics
⇨ according to IST proposals

State-of-the-art

OPTIMIST

☺ In transmission systems
☺ In optical networks

⇨A key concept : GRANULARITY

Technological alternatives :

DWDM versus TDM

Optics Versus Electronics (MUX, Routing)

OPTIMIST

TRANSMISSION

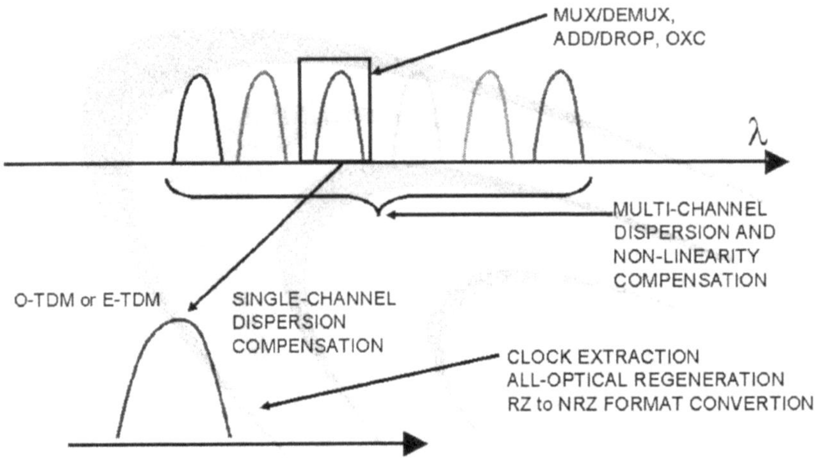

NOIS COST P2 Workshop 26 October 2000

OPTIMIST

Optical Bandwidth

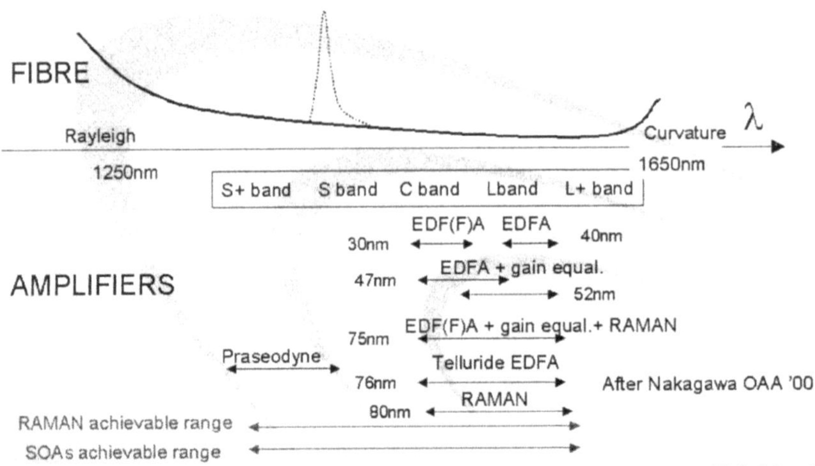

NOIS COST P2 Workshop 26 October 2000

Fibre 's design

OPTIMIST

VAST VARIETIES OF NEW FIBRES WITH TAILORED CHARACTERISTICS

☺Dispersion shifted

☺ Dispersion flattened

☺ Enhanced/reduced non-linearirities

⇨(large effective mode area fibre)

☺ Compensation fibre

⇨(including Reverse dispersion fibre: simultaneous D, D')

NOIS COST P2 Workshop 26 October 2000

Guiness's book of record

(after ECOC '2000 post-deadline papers)

OPTIMIST

☺ DWDM

⇨6.4 Tb/s =160 x 40 Gb/s (@50GHz, 100GHz AWG) over 186 km

⇨5.12 Tb/s =128 x 40 Gb/s (@50GHz/75GHz) over 300 km

⇨7 Tb/s =2x88 x 40 Gb/s (@100GHz/2) over 50 km Bi-direct.

⇨1.28 Tb/s =32 x 40 Gb/s (@100Hz) over 1000 km (6spans)

⇨2 Tb/s =100 x 20 Gb/s (@35GHz, 70GHz) over 2700 km

⇨2.1 Tb/s =211 x 10 Gb/s (@37.5 GHz, FEC) over 7221 km (80km spacing)

☺OTDM

⇨1.28 Tb/s single channel 200 fs pulses over 70 km

- *(3rd + 4th order dispersion compensation)*

NOIS COST P2 Workshop 26 October 2000

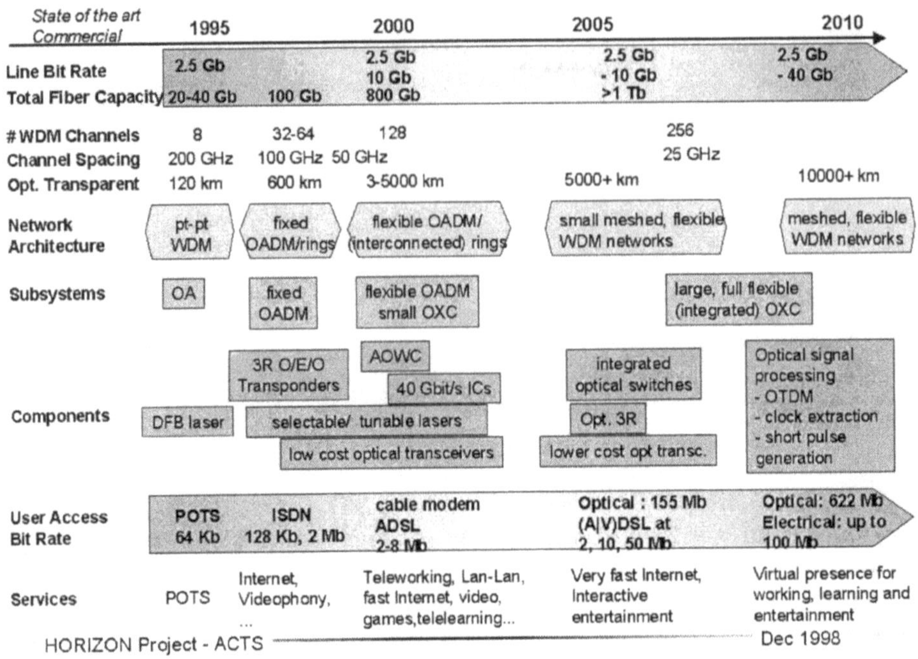
State of the art
Commercial
1995
2000
2005
2010
Line Bit Rate
Total Fiber Capacity
2.5 Gb
20-40 Gb
100 Gb
2.5 Gb
10 Gb
800 Gb
2.5 Gb
- 10 Gb
>1 Tb
2.5 Gb
- 40 Gb
WDM Channels
8
32-64
128
256
Channel Spacing
200 GHz
100 GHz 50 GHz
25 GHz
Opt. Transparent
120 km
600 km
3-5000 km
5000+ km
10000+ km
Network
Architecture
pt-pt
WDM
fixed
OADM/rings
flexible OADM/
(interconnected) rings
small meshed, flexible
WDM networks
meshed, flexible
WDM networks
Subsystems
OA
fixed
OADM
flexible OADM
small OXC
large, full flexible
(integrated) OXC
Components
3R O/E/O
Transponders
AOWC
40 Gbit/s ICs
integrated
optical switches
Optical signal
processing
- OTDM
- clock extraction
- short pulse
generation
DFB laser
selectable/ tunable lasers
Opt. 3R
low cost optical transceivers
lower cost opt transc.
User Access
Bit Rate
POTS
64 Kb
ISDN
128 Kb, 2 Mb
cable modem
ADSL
2-8 Mb
Optical : 155 Mb
(A|V)DSL at
2, 10, 50 Mb
Optical: 622 Mb
Electrical: up to
100 Mb
Services
POTS
Internet,
Videophony,
...
Teleworking, Lan-Lan,
fast Internet, video,
games,telelearning...
Very fast Internet,
Interactive
entertainment
Virtual presence for
working, learning and
entertainment
HORIZON Project - ACTS
Dec 1998

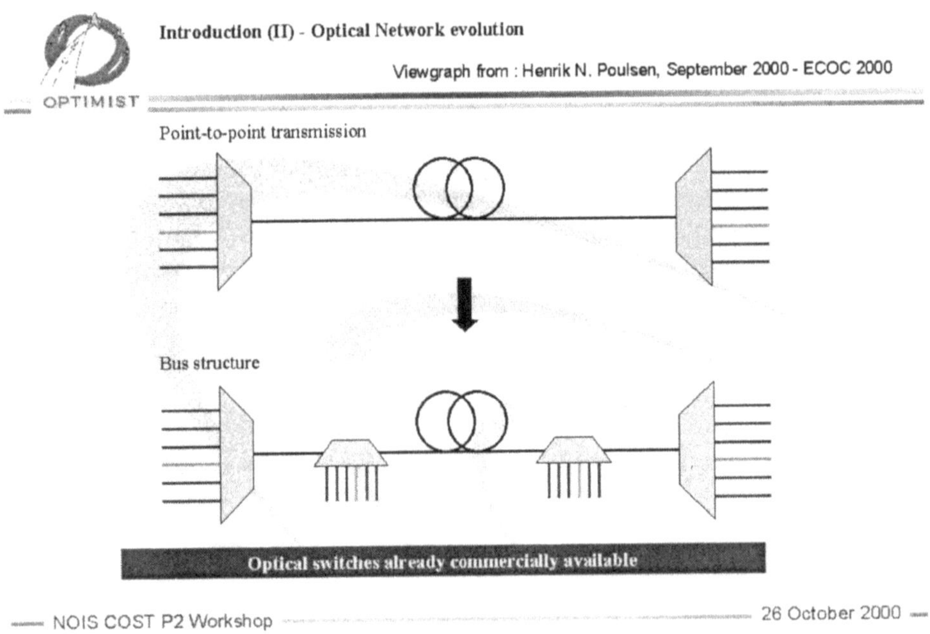
OPTIMIST
Introduction (II) - Optical Network evolution
Viewgraph from : Henrik N. Poulsen, September 2000 - ECOC 2000
Point-to-point transmission
Bus structure
Optical switches already commercially available
NOIS COST P2 Workshop
26 October 2000

OPTIMIST

Network

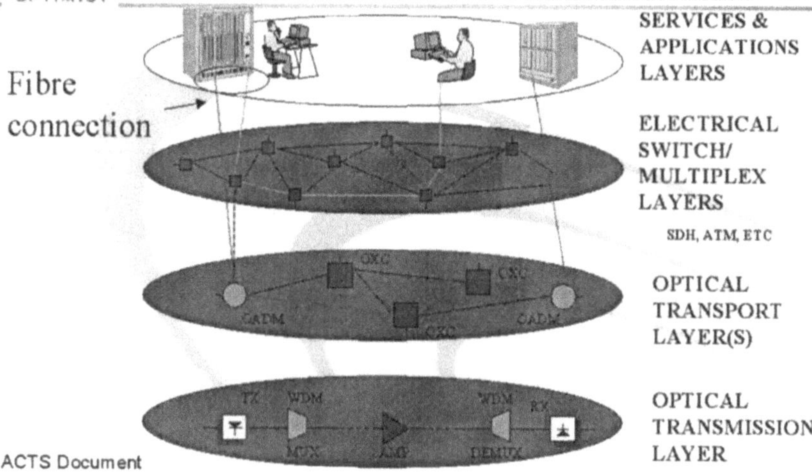

ACTS Document

OPTIMIST

KEOPS Packet Switch

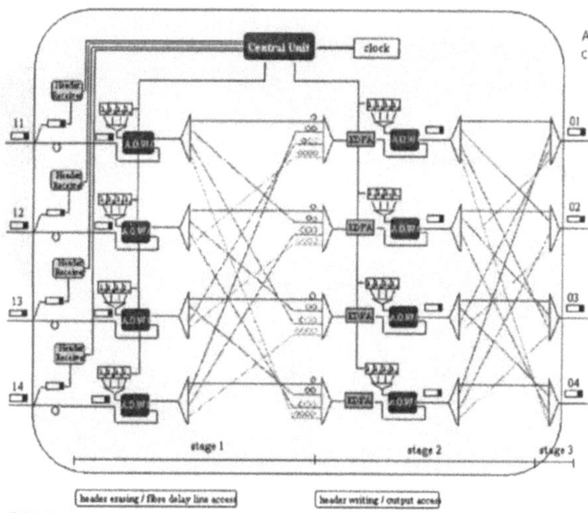

All optical wavelength converter

OPTIMIST

Device demand

- ☺ Mux/ Demux (WDM, TDM)
- ☺ Wavelength converter
- ☺ 3R regenerators (single-, multi - channel)
- ☺ OADM, OXC (circuit switching, packet switching)
- ☺ Active compensators : PMD, GVD
- ☺ Logic gates (?)

NOIS COST P2 Workshop 26 October 2000

OPTIMIST

Mature proposals for non-linear application in optical systems

- ☺ Non-linear operation based on
  - ⇨EA modulators
  - ⇨SOAs
  - ⇨fibres non-linearities

NOIS COST P2 Workshop 26 October 2000

OPTIMIST

Techniques for High Speed Signal Processing (I)

Henrik N. Poulsen, September 2000 - ECOC 2000

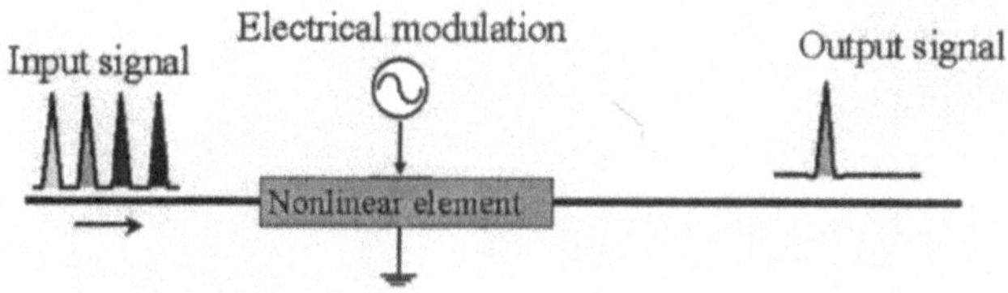

- •EO process
- •Semiconductor based

- **320 Gbit/s clock recovery and 320 to 10 Gbit/s demultiplexing**
  B. Mikkelsen et al., Post deadline, OFC, Baltimore 2000

OPTIMIST

Demultiplexing - Electro-absorption modulator (I)

Henrik N. Poulsen, September 2000 - ECOC 2000

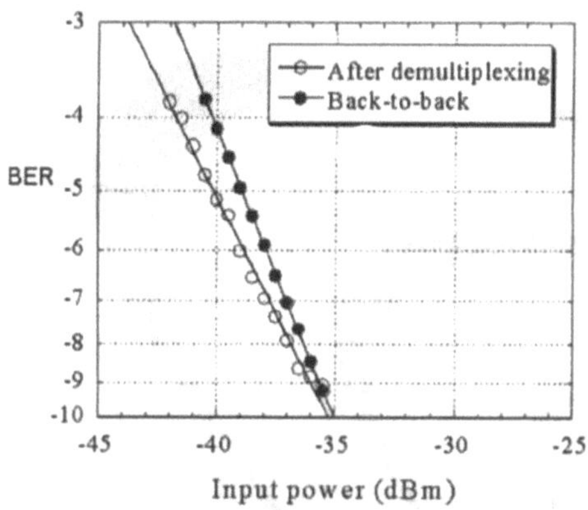

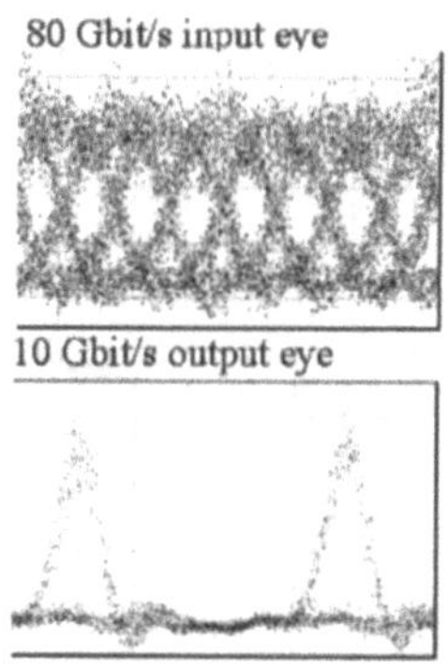

80 to 10 Gbit/s demultiplexing is pushing the limit ⇒
more complex setup necessary for higher bit rates

OPTIMIST

Non-linearities in fibres and SOAs

☺ **Fibres**

- ⇨Kerr -> SPM, XPM.
- ⇨Brillouin, Raman

☺**SOAs**

- ⇨Inter-band non-linear behaviour (0.1 ns)
  - *Carrier density fluctuations -> SPM, XPM, SGM, XGM + FWM*
- ⇨Intra-band non-linearities (ps or faster)
  - *Spectral hole burning, Carrier heating, Intra-band abs. -> F□WM*

OPTIMIST

Techniques for High Speed Signal Processing (II)

Henrik N. Poulsen, September 2000 - ECOC 2000

Cross gain (absorption) modulation

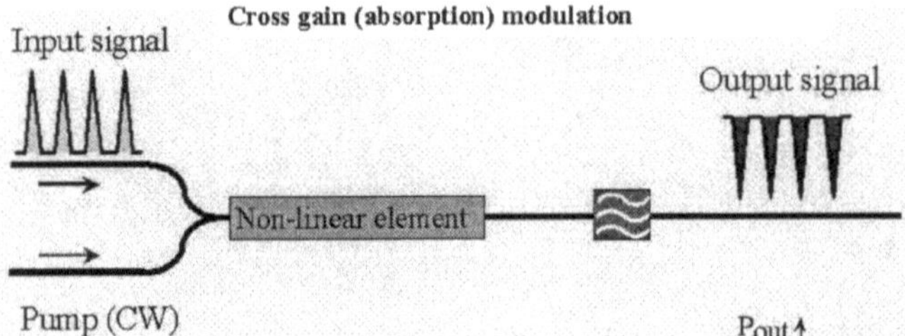

XGM process

- EA based
- **SOA based**

- **100 Gbit/s conversion over 5 nm using specially optimised SOA**

A. D. Ellis et al. Electronics Letters, vol. 34, No. 20, 1998

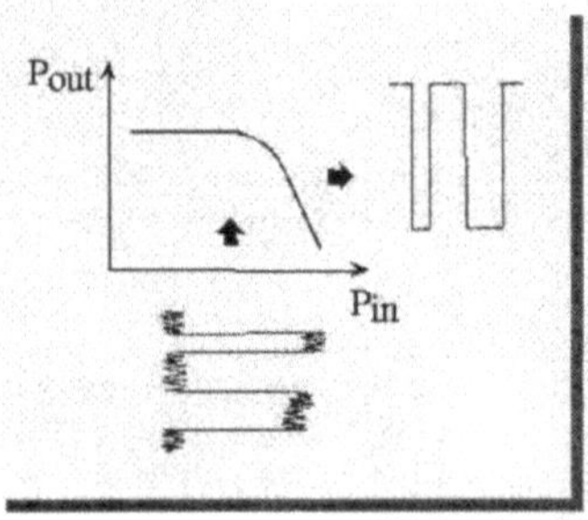

OPTIMIST

Techniques for High Speed Signal Processing (III)

Henrik N. Poulsen, September 2000 · ECOC 2000

Four wave mixing

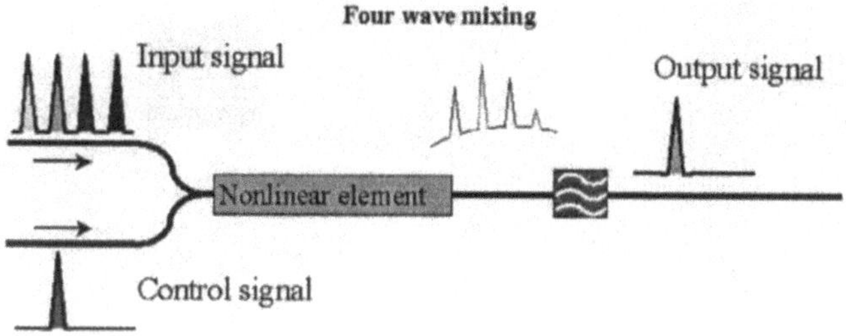

FWM process

- Fiber based
- SOA based

- FWM in fiber: ***500 Gb/s demux, Morioka et al., NTT, 1996***
- FWM in SOA: ***400 Gb/s phase comp., Kawanishi et al., NTT, 1996***

UPGRADE

OTDM performance of TOAD

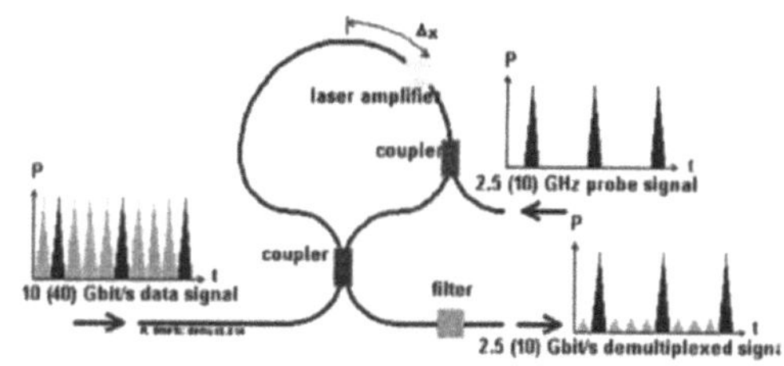

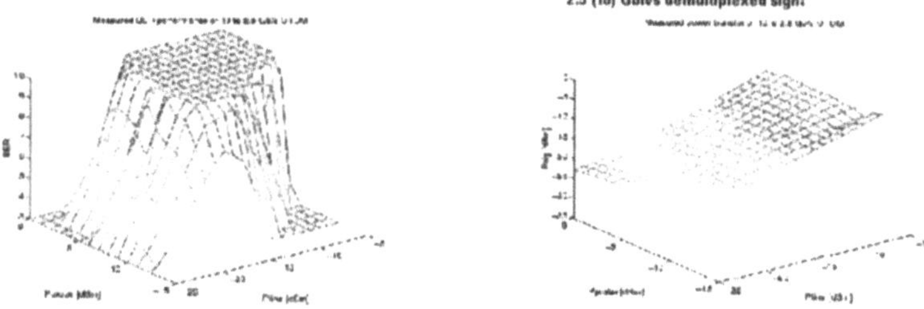

Philips
Optoelectronics
Research

Interferometric structure - fiber based

Henrik N. Poulsen, September 2000 · ECOC 2000

OPTIMIST

640 to 10 Gbit/s demultiplexing

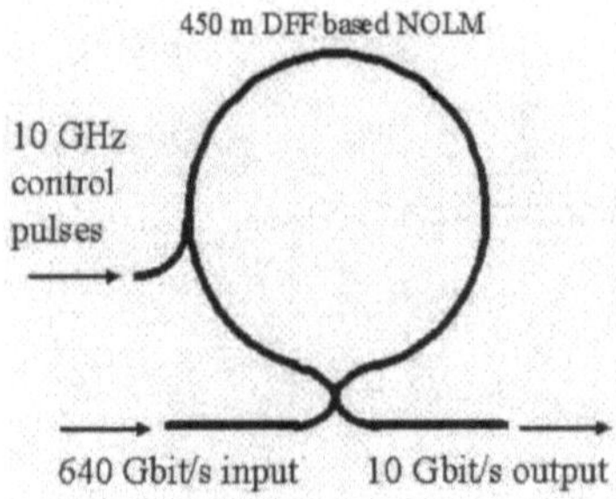

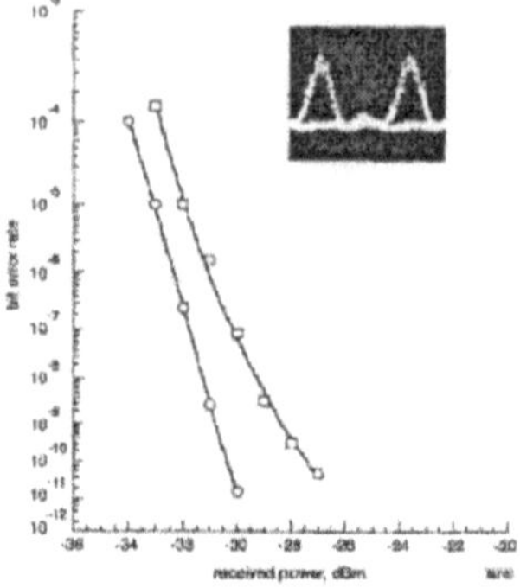

Yamamoto et al., NTT, Electronics Letters, Vol. 34, No. 10, 1998

Error free demultiplexing performed using fiber non-linearity in NOLM
- *very effective; however bulky and difficult to stabilise*

Multi-channel wavelength conversion

OPTIMIST

Broadband wavelength conversion using SHORT HIGH-NONLINEARITY non-polarization maintaining FIBRE

23.3 nm conversion band
200 m HWHM Fibre

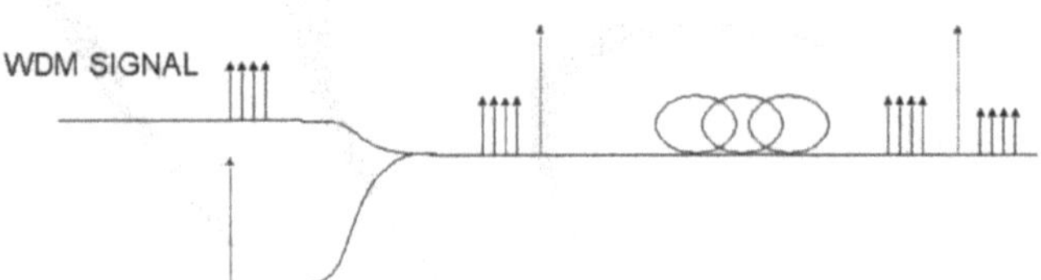

Azo et al ECOC '99 PP.II-226

OPTIMIST

Demultiplexing - Mach-Zehnder Interferometer

Henrik N. Poulsen, September 2000 - ECOC 2000

40, 80 and 160 to 10 Gbit/s demultiplexing

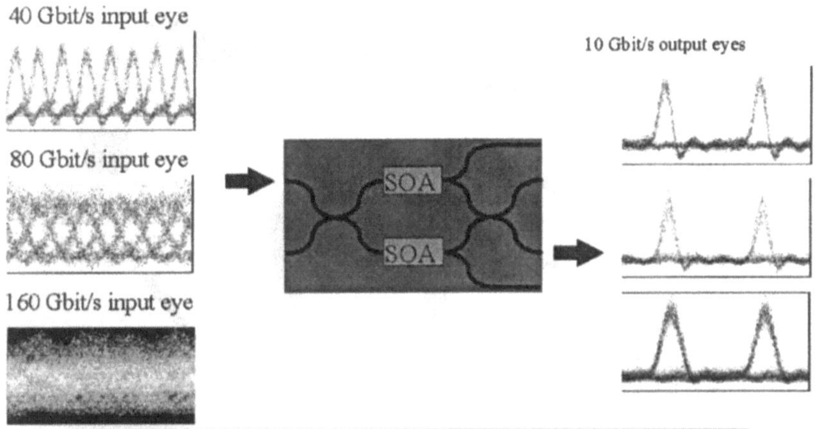

No significant pulse distortions observed after demultiplexing

NOIS COST P2 Workshop — 26 October 2000

OPTIMIST

Add and drop multiplexing - results @ 40 Gbit/s

Henrik N. Poulsen, September 2000 - ECOC 2000

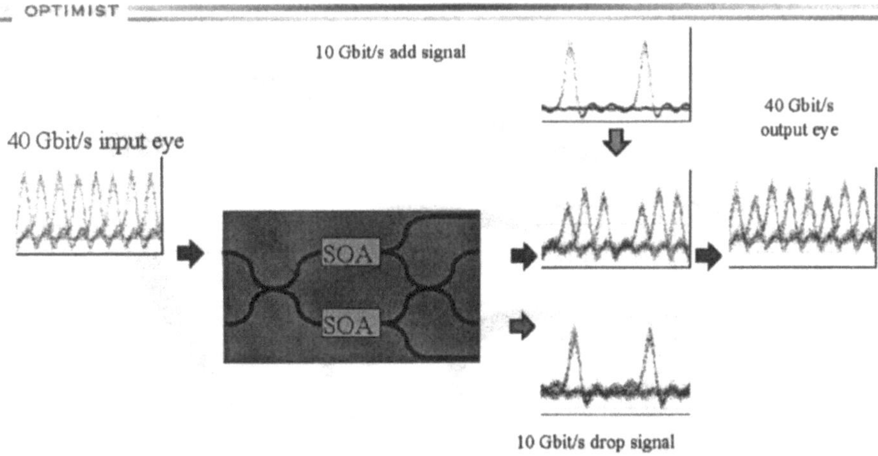

Simple and compact realisation of all-optical add and drop multiplexer

NOIS COST P2 Workshop — 26 October 2000

OPTIMIST

RZ-to-NRZ format conversion - eye-diagrams

Henrik N. Poulsen, September 2000 - ECOC 2000

RZ-output signal | **NRZ-output signal**

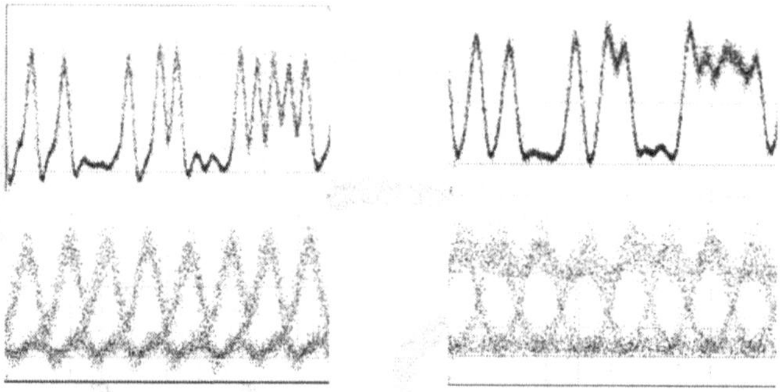

Conversion to RZ and NRZ possible by adjusting delay

RZ conversion: pulse width increases from ~8 ps to ~12 ps (=> fast!)

NOIS COST P2 Workshop — 26 October 2000

OPTIMIST

OPEN optical cross-connect architecture

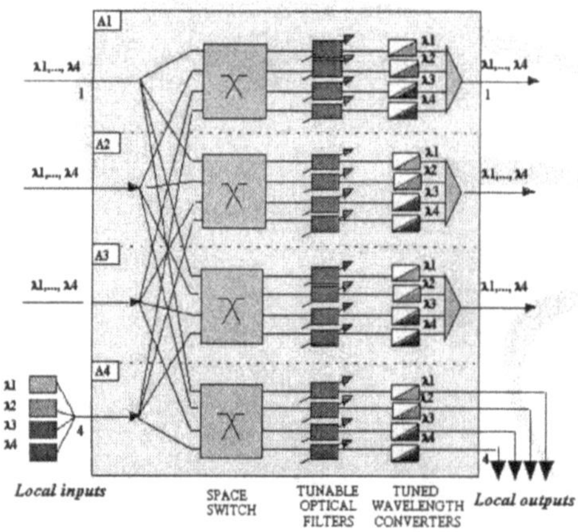

<u>Features</u>

- Broadcast-and-Select architecture
- All-optical technology
- 3 switching functional blocks:
  - WDM space-switching
  - tunable wavelength selection
  - tuned wavelength conversion
- Local input/output ports

NOIS COST P2 Workshop — 26 October 2000

OPTIMIST

New trends (IST proposals)

- ☺ SOA and fibre based signal processing
- ☺ Parallel signal processing
- ☺ Photonic crystals
- ☺ New materials
  - ⇨Semiconductors
  - ⇨Polymers (Integrated optics, fibres,...)
- ☺ Quantum Communications

NOIS COST P2 Workshop — 26 October 2000

OPTIMIST

Optical signal processing

DO-ALL Digital Optical Fibre Logic Modules
- All optical bit error rate measurement @ 40Gb/s
- An optically addressable, exchange-bypass switch @ 40Gb/s

PIANOS : Processing o information by arrays of non-linear optical solitons
Vertical cavity semiconductor resonators which sustain spatial solitons; informatic applications.
Purely optical processor based on bistability and mobility of optical solitons

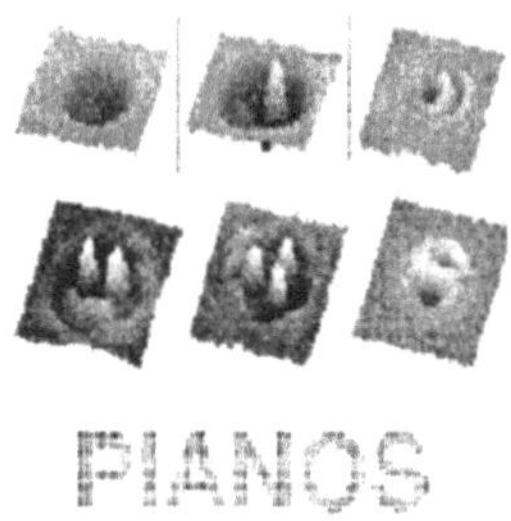

NOIS COST P2 Workshop — 26 October 2000

Optical signal processing

OPTIMIST

ANLM-OTDM : Active non-linear Mirrors For Optical Time Division Multiplexing.
Second-order process in microstructures (micro-cavities and photonic band-gap structure).

High speed above 40Gbs only limited by pulse width

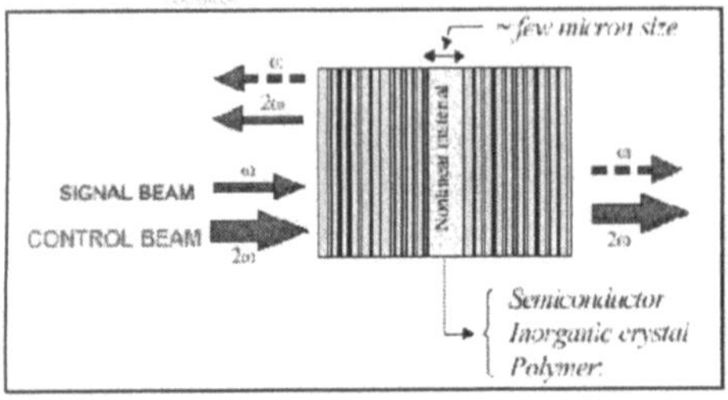

The control beam at frequency 2ω modifies the reflectance for the signal beam at frequency ω

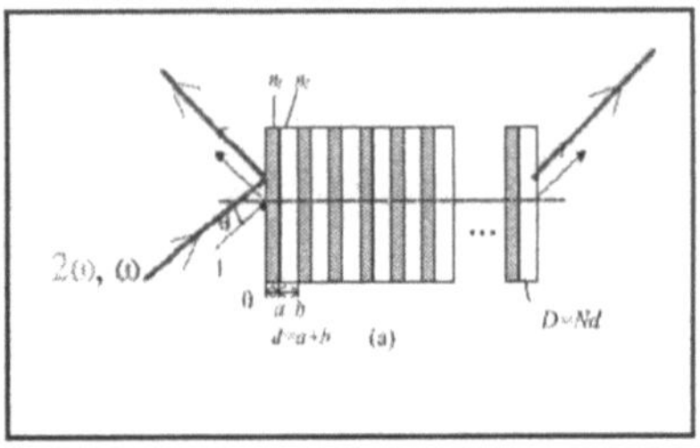

Resonance with the maxima of the density of modes of the PBG structure increases the nonlinear interaction

Photonic Crystal

OPTIMIST

PCIC : Photonic Crystal Integrated Circuits
Applied the concept of 2D photonic Crystal
To PICs on InP

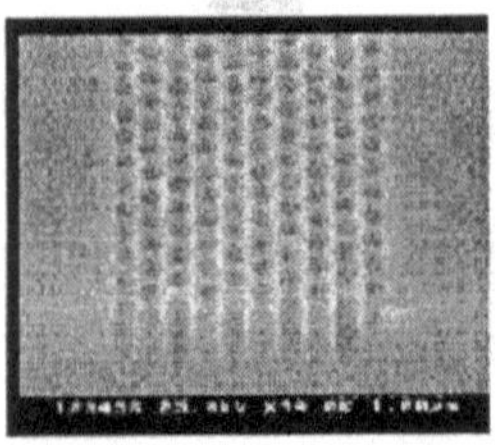

Photonic Crystal Waveguide

PICCO : Photonic Integrated Circuit using Photonic Crystal Optics
WDM components based on dispersion engineering in PC. Monolithic parametric wavelength converter

PHOBOS : Photonic Crystal Based on Opal Structure.
Real 3D Photonic cristals

OPTIMIST

New materials

Amber blue emitters targeting high temperature operation.
GaN based LED **AGETHA**

« IN VEHICLE POF NETWORK »

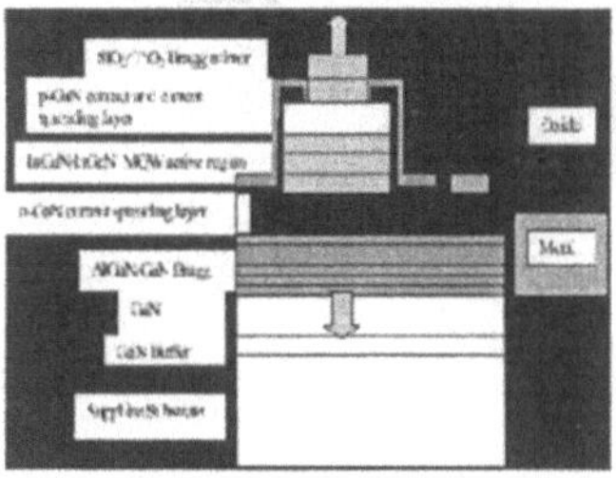

GaAs-based emitters for fibre-optical data and telecommunication **GIFT**
GaAs-based long wavelength emitters (1300nm/1550nm) using new quantum well structures

GaAs-second window quantum dot lasers

LOBSTER : Large optic bandwidth by amplifier systems based on Telluride fibres doped with rare earths

OPTIMIST

Quantum communications

EQUIS : Enabling technology for quantum transmission systems
New components ans sub-system technology for manufacture at relatively low cost quantum cryptography.

QUCOM : Long distance photonic quantum communication

- User friendly source of complex quantum states, photon twins, triplets...
- Non-linear optical crystals ...
- Deployment of novel sources in quantum comm. protocols.
- Cryptography
- Quantum error correction for transfer of quantum information through noisy channels.

OPTIMIST

Quantum cryptography

1 photon sources (as an example of Q-bit) for quantum key exchange

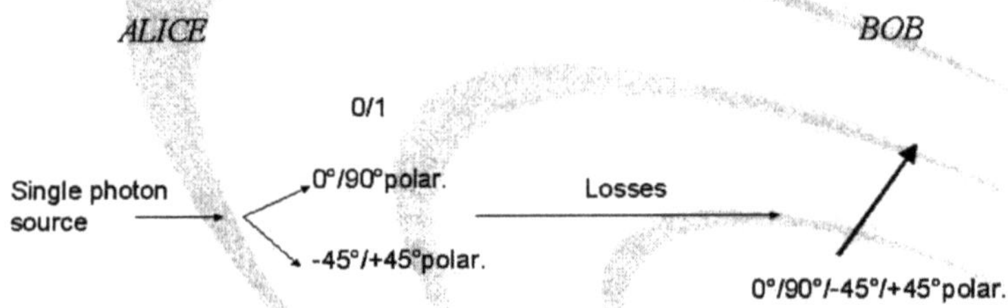

Photon lost = no information
Wrong set of polariser = no information
Proper set of polariser = information

Intruder = increased BER (on a recognised message)

OPTIMIST

Summary

☺ Ever growing bandwidth on offer for optical transmission systems and optical networks

☺ Electronic limits -> alternative OTDM and DWDM+ETDM

☺ Requirement for devices in the routing process.

☺ Non-linear application already proposed : main challenger SOAs and Non-linear fibres.

☺ New candidates and new requirement.

☺ " The new generation optical communications networks are being built using devices that emerged from fundamental research from the last ten years... Pursuing this research into basic physics may lead to a completely new area of communication and computing, with perspectives that we are not likely to imagine today " (OPTIMIST)

WAVELENGTH CONVERSION, OPTICAL PARAMETRIC AMPLIFICATION AND OSCILLATION IN PERIODICALLY POLED TI:LINBO$_3$ OPTICAL WAVEGUIDES

W. SOHLER, D. HOFMANN, G. SCHREIBER
Universität Paderborn, Angewandte Physik
Warburger Str. 100, D-33098 Paderborn/Germany
Phone: +49 5251 60-2712, Fax: +49 5251 60-3422
Email: sohler@physik.uni-paderborn.de

Abstract

Quasi-phase matched quadratic nonlinear waveguide devices in periodically poled lithium niobate (PPLN) are reviewed with emphasis on Ti:(Er:)PPLN structures. In particular, difference frequency generators, optical parametric amplifiers and oscillators and combinations of lasers and nonlinear frequency converters are presented.

1. Introduction

The recent demonstration of periodically poled optical waveguides in $LiNbO_3$ allows the development of very efficient quasi-phase matched second harmonic and difference frequency generators (wavelength converters), parametric amplifiers, all-optical ultra-fast switches, optical parametric oscillators and pulse compressors. Waveguide design and periodicity of the domain pattern determine the phase match wavelengths in the broad range from the visible to the mid-infrared ($\lambda \approx 4\mu m$). Moreover, Ti:$LiNbO_3$ devices can be combined with rare-earth- (in particular: Er-) doped amplifiers and lasers even in the same waveguide structure. There are numerous applications in the fields of optical communications, of all-optical signal processing and of spectroscopic analysis and monitoring of e.g. the environment. It is the aim of this contribution to review quasi-phase matched quadratic nonlinear waveguide devices in periodically poled lithium niobate (PPLN) with emphasis on Ti:(Er:) PPLN structures.

2. Periodically Poled Ti:(Er:) LiNbO$_3$ Waveguides

To develop nonlinear integrated optical devices mainly proton exchanged $LiNbO_3$ waveguides were used up to now [1]. Recently, also the fabrication of up to 80mm long Ti:PPLN waveguides with periodicities around 17μm and 32μm was reported [2]. Even Ti:Er:$LiNbO_3$ waveguides, fabricated in an Er-diffusion-doped substrate, could be poled

A. Driessen (ed.), Nonlinear Optics for the Information Society, 23–27.

using the electric field technique. The domain pattern was investigated by chemically etching the waveguide surface revealing a nearly ideal 1:1 duty cycle (see Fig. 1). Moreover, the domain structure was studied by second harmonic generation- (SHG-) microscopy demonstrating a thickness of the domain walls below the resolution limit of the microscope.

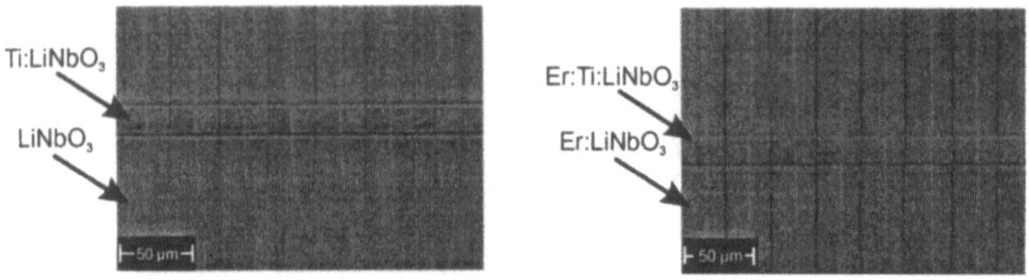

Figure 1: Chemically etched surface of a periodically poled undoped (left) and Er-diffusion-doped (right) Ti:$LiNbO_3$ channel guide. The periodicity is 32µm.

The short period waveguides were designed as single-mode channels around λ = 1550nm for near-infrared (NIR) nonlinear interactions. They were tested as second harmonic generators using a tunable, single-frequency semiconductor laser as pump source. Measured and calculated characteristics (phase-matching curves) nearly coincide demonstrating the excellent homogeneity of the periodically poled waveguides. Their high quality with losses as low as 0.15dB/cm results in a normalized efficiency of 442 % W^{-1} (50mm long waveguide) or of 18 % $W^{-1}cm^{-2}$ [2].

The long period guides were single-mode channels around λ = 3400nm to allow mid-infrared (MIR) nonlinear interactions. The losses are as low as 0.03dB/cm, ideal for resonant structures.

3. Wavelength Conversion (DFG) and Optical Parametric Amplification (OPA)

Waveguides of both periodicities were investigated as difference frequency generators (DFG).

The short period guides were studied as wavelength converters for the 1.5µm communication band using a Ti:sapphire laser as pump source ($\lambda_p \approx$ 780nm) and a tunable semiconductor laser as signal source (1500nm < λ_s < 1580nm). With 50mm long waveguides a device efficiency of 318 % W^{-1} or 13 % $W^{-1}cm^{-2}$ was measured [2].

Instead of a Ti:sapphire laser with all the problems to selectively excite a single pump mode at $\lambda_p \approx$ 780nm, a narrow linewidth semiconductor laser of $\lambda_f \approx$ 1560nm can be used, if necessary in combination with a high power erbium-doped fiber amplifier (EDFA). In this case the pump wave for DFG is generated inside the device by SHG; coupling to a single SHG-mode is guaranteed by the phase-match condition. In this "cascaded" process the overall DFG-efficiency is somewhat reduced. Nevertheless, -14dB or 4 % has been experimentally obtained as device efficiency in a 80mm long periodically poled Ti:$LiNbO_3$ waveguide at a coupled pump power of 155mW. The spectral bandwidth for wavelength conversion (at fixed fundamental and pump wavelengths) is 55nm in good agreement with theoretical predictions (see Fig. 2) [3].

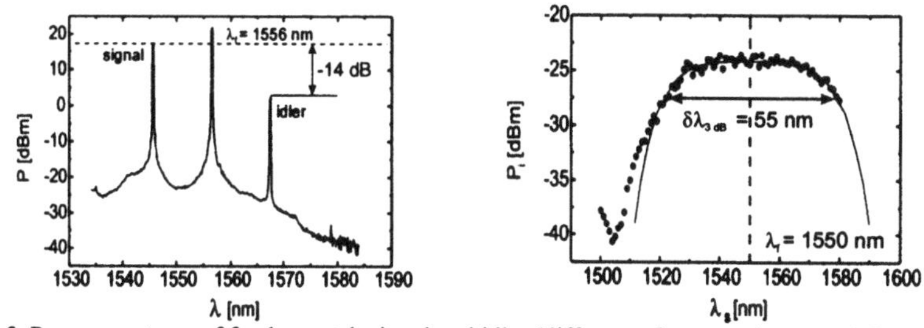

Figure 2: Power spectrum of fundamental, signal and idler (difference frequency) waves (left); generated idler power versus signal wavelength (right).

DFG is always accompanied by optical parametric amplification (OPA) of the signal wave, as pump photons decay into one idler and one signal photon simultaneously (energy conservation). However, to be useful e.g. in optical communications, much higher efficiencies and therefore higher pump power levels are required.

Recently, with ~500mW coupled fundamental power the efficiency for cascaded DFG was enhanced to –1.8dB accompanied by an amplification of the signal wave of +2dB. Theory even predicts a gain of +14dB in a 80mm long "cascaded" OPA at the same power level; in a "direct" converter the gain should grow up to +19dB.

The $\chi^{(2)}$-nonlinear processes are extremely fast paving the way to all-optical switching. This was demonstrated using a harmonically modelocked Ti:Er:$LiNbO_3$ waveguide laser [4] as fundamental source (λ_f = 1561nm) for "cascaded" DFG/OPA. With fundamental pulses of ~6ps halfwidth somewhat short idler pulses were generated with an efficiency of 1.8%; in this way e.g. all-optical time division (de-)multiplexing (TDM) in combination with wavelength division (de-)multiplexing (WDM) can be achieved.

The long period waveguides were used as MIR-difference frequency generators mixing the radiation of a tunable extended cavity semiconductor laser (1500nm < λ_p< 1580nm), amplified by an EDFA, and of a mid-infrared He-Ne laser (λ_s = 3391nm) to generate the difference frequency (idler) radiation of a wavelength around λ_i = 2800nm. Using 80mm long waveguides a device efficiency of 105 % W^{-1} was achieved, by far the best result ever reported [5]. Such devices are ideally suited to develop compact efficient MIR-DFG spectrometers using semiconductor lasers as pump and signal sources.

4. Optical Parametric Fluorescence (OPF) and Oscillation (OPO)

Waveguides of both types as described above were also used to demonstrate the single-pass generation of spontaneous and stimulated parametric fluorescence. This fluorescence is induced by the quantum-mechanical zero point (vacuum) fluctuations of the electromagnetic field. Spontaneous optical parametric fluorescence (OPF) is an ideal means to measure the tuning characteristics of three wave $\chi^{(2)}$-processes. A tuning range of about 750nm (750nm) has been achieved in the NIR- (MIR-) spectral range by a variation of the pump wavelength. At higher pump power levels, stimulated PFG is obtained. Using a Q-switched Ti:Er:$LiNbO_3$ waveguide laser [4] with 5ns long pulses of about 500W peak power (λ_p = 1562nm) MIR-pulses (λ_s = 2930nm; $\lambda_i \approx$ 3300nm) were generated with a peak power of up to 10mW.

Spontaneous parametric fluorescence represents the noise which starts optical parametric oscillation (OPO) in nonlinear waveguide cavities, if the pump power exceeds a certain threshold. Such (doubly resonant) OPOs have been developed for both the NIR- and MIR-spectral ranges by coating the end faces of waveguides of both types with dielectric multilayer mirrors (SiO_2/TiO_2). (for the MIR-devices commercial mirrors were used pressed against the waveguide end faces.) Due to the low losses of Ti:$LiNbO_3$ waveguides – especially in the MIR – high Q cavities could be fabricated leading to a low oscillation threshold of about 35mW of the NIR-OPO and of only 14mW of the MIR-OPO [6] (see Fig. 3, left). At higher pump power levels the OPO output power was increased up to several mW in cw-operation (see also Fig. 3, left). By optimizing the resonator design much higher output levels can be expected.

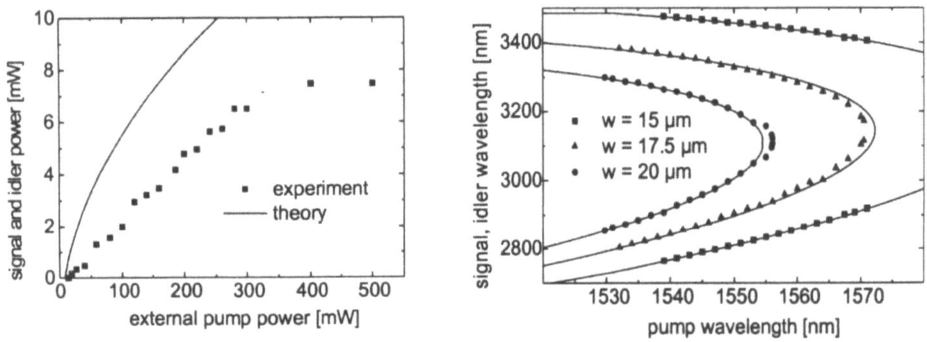

Figure 3: Power characteristics (left) and tuning characteristics (right) of 80mm long Ti:$LiNbO_3$ MIR-OPOs of different waveguide width w and of Λ = 31,6µm domain periodicity.

The spectral tuning width of the OPOs was mainly determined by the width of the reflectivity band of the dielectric mirrors. In the NIR-device it extends from 1400nm to 1800nm, in the MIR-device from 2700nm to 3500nm. The tuning characteristic is not only determined by the periodicity of the ferroelectric domain pattern, but also by the waveguide width, which modifies the effective indices and in this way the phase match wavelengths (see also Fig. 3, right). The fine tuning behavior of the doubly resonant OPOs shows the well-known "saw-tooth" behavior according to the requirements of energy conservation, phase matching and simultaneous resonance of signal and idler waves. Singly resonant OPOs are the better choice for high resolution spectroscopy; corresponding devices are currently developed.

5. Laser/Frequency Converter Combinations

As laser-active Ti:Er:LiNbO$_3$ waveguides can be periodically poled as well (see Fig. 1, right), integrated optical lasers (or amplifiers) and quasi-phase matched nonlinear frequency converters can be combined not only on the same substrate, but even in the same waveguide structure. This idea leads to a number of attractive concepts, only partly realized up to now.

The most straightforward device is a self-frequency doubling laser. A high conversion efficiency can be expected, if the resonator is designed to maximize the enhancement of the laser field. Even a doubly resonant device (resonant at laser and second harmonic wavelengths) should be possible. First experimental results of a self-frequency doubling laser have been reported recently [7].

The same principle can be applied to fabricate an amplifying difference frequency generator in a periodically poled Ti:Er:LiNbO$_3$ waveguide. By pumping the Er-ions a wavelength-dependent single-pass amplification up to 2dB/cm should be possible. Also this concept has been experimentally demonstrated recently.

Furthermore, a MIR-OPO can be designed with an intracavity Er-laser as pump. Currently, experiments are going on to realize such a device.

6. Conclusions

A variety of efficient nonlinear quasi-phase matched waveguide devices has been demonstrated with PPLN as substrate material. Now an optimization for specific applications is required to exploit the great potential of nonlinear integrated optics in optical communications, all-optical signal processing and spectroscopic analysis and monitoring.

References

[1] see e.g.: M.H. Chou, I. Brener, K. Parameswaran, and M.M. Fejer, *Proc. ECIO '99*; Torino, p. 497 (invited)

[2] G. Schreiber, R. Ricken, and W. Sohler; *Proc. ECIO '99*, Torino, p. 95

[3] G. Schreiber, D. Hofmann, W. Grundkötter, R. Ricken, and W. Sohler; *Proc. NLGW '99*, Dijon, p. 468

[4] S. Balsamo, Ch. Becker, I. Montrosset, Th. Oesselke, J. Pandavenes, R. Ricken, K. Rochhausen, G. Schreiber, D. Sciancalepore, W. Sohler, H. Suche, and R. Wessel; to appear in *IEEE J. Selected Topics Quantum Electron., special issue on Integrated Optics* (invited)

[5] D. Hofmann, G. Schreiber, C. Haase, H. Herrmann, W. Grundkötter, R. Ricken, and W. Sohler; *Opt. Lett. 24* (1999) 896

[6] D. Hofmann, H. Herrmann, G. Schreiber, C. Haase, W. Grundkötter, R. Ricken, W. Sohler; *Proc. NLGW'99*, Dijon, p. 465 (invited)

[7] G. Schreiber, K. Rochhausen, R. Ricken, and W. Sohler; *Proc. LASER '99*, Munich, p. 147

ENHANCED NONLINEAR OPTICAL RESPONSE OF LIQUID CRYSTALS

F.SIMONI, O.FRANCESCANGELI, D.E.LUCCHETTA,
L.LUCCHETTI
Dipartimento di Fisica e Ingegneria dei Materiali e del Territorio e INFM,
Università di Ancona
Via Brecce Bianche, 60131 Ancona, Italy

Abstract

An account of the most relevant phenomena leading to the enhancement of the nonlinear optical response of liquid crystals is reported. The giant nonlinearity due to collective reorientation can be increased by two order of magnitude through dye-doping, leading to a nonlinear refractive index n_2 of the order of 10^{-3} cm^2/W. Recently this value has been overcome in in Methyl Red-doped liquid crystals where $n_2 > 1$ cm^2/W have been measured. We show that this extraordinarily large nonlinear response can be also induced by short pulse excitation.

Molecular reorientation is responsible for the nonlinear optical response of many liquids made by anisotropic molecules. Among them liquid crystals (LCs) have the peculiar feature to keep the anisotropy on a macroscopic scale due to the long range molecular correlation; as a consequence they show a collective behavior that enhances the molecular reorientation and leads to the so-called "giant nonlinearity" (GON)[1-3]. This denomination is easily understood when comparing the nonlinear dielectric constant of a nonlinear isotropic liquid ($\varepsilon_2 = 10^{-10}$ esu) to the one obtainable in a pure nematic liquid crystal ($\varepsilon_2 > 10^{-2}$ esu). The collective character of LCs make useful to treat these phenomenon under a macroscopic approach making use of the concept of molecular director $\vec{\delta}$. This is a unit vector representing the average direction of the molecular long axis. In this way the collective properties are well described by a continuum theory where the macroscopic properties of the material vary according to the local orientation of $\vec{\delta}$. Since this orientation determines the optical properties of the medium and the electric field of the light can reorient $\vec{\delta}$, a nonlinear respose is easily foreseen, and it is enhanced in an amazing way by the macroscopic anisotropy of the material.
The extraordinary index is usually written as

A. Driessen (ed.), Nonlinear Optics for the Information Society, 29–33.

$$n(\theta) = \frac{n_{//} n_{\perp}}{\sqrt{n_{//}^2 \cos^2(\theta) + n_{\perp}^2 \sin^2(\theta)}}$$

where $n_{//}$ and $n_{\perp}$ are the refractive index correspondent to directions parallel and perpendicular to $\vec{\delta}$, while θ is the angle between the light wavector and $\vec{\delta}$. Any light-induced reorientation of the director change θ and as a consequence the refractive index of the extraordinary wave. Far from saturation, this refractive index can be written in the usual way as:

$$n = n(0) + n_2 I$$

In pure liquid crystals the "optical torque" is due to the orienting effect of the electric field of the wave, which acts in a way similar to a static field. The optical torque is given by:

$$\vec{\tau}_{opt} = \frac{\Delta\varepsilon}{8\pi}\left(\vec{\delta}\cdot\vec{E}^*\right)\left(\vec{\delta}x\vec{E}\right)$$

A standard treatment through minimization of the free energy of the system allows to calculate the equilibrium orientation of the director at a given intensity impinging on the sample. As a further step, from this orientation it is possible get the nonlinear refractive index. Such an approach accounts for the main features observed in the experiments and explain the occurrence of the measured nonlinear refractive index n_2 which reach values of the order of 10^{-5} cm^2/W. This giant nonlinear optical response allows the observation of many effects like self-focusing[4], self-phase modulation[5], four-wave mixing[6] with c.w. light beam of moderate power.
In 1990 Janossy and coworkers discovered[7-8] that doping a nematic liquid crystal by a small amount of dye of the family of anthraquinone derivatives could lead to an enhancement of the optical reorientation effect by more than two order of magnitude. All the experiments performed to investigate this phenomenon have shown that it is possible to introduce[9] a phenomenological enhancement factor ζ in the expression of the optical torque. In this way the total optical torque can be expressed by:

$$\vec{\tau}'_{opt} = \vec{\tau}_{opt} + \vec{\tau}_{ph}$$

where τ_{ph} is the additional torque due to the dye absorption:

$$\vec{\tau}_{ph} = \frac{\zeta}{8\pi}\left(\vec{\delta}\cdot\vec{E}^*\right)\left(\vec{\delta}x\vec{E}\right)$$

The use of this expression for the photo-induced torque means the assumption of a local mechanism of enhancement , thus excluding effects based on thermal or mass gradient, photo-refraction, and so on. The present models which explain the Janossy's effect are based on the photo-induced variation of the microscopic interaction between the dye molecules and the LC molecules. The basic point is that this interaction is different if the dye molecule is in the ground state or in the excited state. In the limit of small dye anisotropy and far from saturation Marrucci and Paparo[10] found an approximate expression for the enhancement parameter as:

$$\zeta = \frac{16\pi}{15}\sigma N_d S \frac{\tau_e N_h}{1+6D_e\tau_{ec}}\left(u_{eh} - \frac{D_e}{D_g}u_{gh}\right)$$

where S is the order parameter of LC, u_{eh} and u_{gh} are coefficients which describe the strength of the dye-host orientational interaction, D_e and D_h are the rotational diffusion constant of the dye molecules in the excited and in the ground state respectively, τ_e is the lifetime of the excited level, N_d and N_{LC} are the number densities of dye and LC, and σ is the absorption cross section. Accordingly the term in parenthesis is the key one pointing out that the effect is due to a change of the intermolecular interaction and of the rotational diffusion constant.

A different family of dyes may have an additional reorienting effect on LCs: the azo-dyes. Upon light absorption they undergo a conformational change which is known as photo-isomerization. The photo-isomerization process begins with absorption by a molecule in the trans state (energetically more stable), followed by a non radiative decay to the ground state in either the cis or trans form. From the cis state (metastable) the molecule can thermally decay to the trans state. The different shape of the molecules in the two states (elongated in the trans state, V-shaped in the cis state) account for the mentioned change in the interaction with the LC molecules. In other words in this case we may apply the same model used for anthraquinone derivatives just considering two different species in the excited state. In this way we may have two different contributions to the enhancement factor : ζ_T for the trans isomer and ζ_C for the cis isomer. The total enhancement[11] factor will depend on the concentration ratio of the two species:

$$\zeta = \zeta_T + (\zeta_C - \zeta_T)X$$

where X is the cis isomer concentration.

We can notice that the two species act in different way since the trans isomer contributes with a positive term while the cis isomer gives rise to a negative term, this fact is due to the particular shape of the isomer that decreases the orientational order of the mixture. The consequence is usually an overall enhancement much lower than the one obtained with the anthraquinone derivatives.

Therefore we can say that the enhancement of GON due to the photo-induced additional torque of the dye molecules allows to get n_2 of the order of 10^{-3} cm^2/W.

A similar value can be obtained by exploiting a "photorefractive-like" effect. This effect can be observed when spatially modulated light (e.g. the interference pattern of two laser beams) impingies on a nematic liquid crystal sample where a low (1-2 V) d.c. bias electric field is applied[12-13]. The effect is based on three mechanisms[14]: i) optical charge generation and ion diffusion with consequent director reorientation ; ii) optical charge generation and space charge field formation through the dielectric and conductivity anisotropy (Carr-Helfrich effect) and consequent director reorientation ; iii) optical charge generation followed by material flow and consequent director reorientation.

As a result the total torque acting on the molecular director is:

$$\vec{\tau}_{total} = \vec{\tau}_{\Delta\varepsilon} + \vec{\tau}_s + \vec{\tau}_K + \vec{\tau}_{opt}$$

where the four terms on the right correspond respectively to the dielectric, the shear-induced, the elastic and the usual optical torque. The key factor is the dielectric torque

wihch makes the director reorientation strongly dependent on the dielectric anisotropy. This fact accounts for the ehnancement of the nonlinear response observed in this case, if, for instance, we consider the nematic liquid crystal 5CB, the dielectric anisotropy is about 40 times higher than the optical anisotropy. In this material doped by Rhodamine 6G or by fullerene C_{60} , Khoo[15] under the usual two beam coupling geometry have measured nonlinear refractive index n_2 of the order of 10^{-3} cm^2/W.
More recently a huge nonlinear response has been detected[16] in nematic liquid crystals doped by methyl-red (1% in weight). In these samples, using two equal power writing beams expanded to a beam diameter of 1.5 cm, a visible grating diffraction has been observed with a total input power as low as 100μW corresponding to an optical intensity of 40 $\mu W/cm^2$, while a maximum efficiency of 30% has been observed at intensity of 1 mW/cm^2. The measured parameters correspond to $n_2 > 1$ cm^2/W !
The origin of this extraordinarily large nonlinear response is now subject of investigation. Thermal gradient effects are ruled out because of the observed dependence on the light polarization and because of the measured decay time which is consistent with a director reorientation. As a matter of fact a photovoltaic effect is detected when the these sample are irradiated, due to photocharge generation in the methyl red-doped liquid crystal and subsequent diffusion and redistribution of ions with different mobilities. However the measured value of the photo-induced voltage (of the order of 1mV) is too low to explain the nonlinear response through a photorefractive-like effect. Most likely the measured photo-voltage may be an indication of the modifications induced by the light beam on the boundary surfaces which could affect the director orientation through the elasticity of the medium.
It is quite interesting that such a huge nonlinear response is also obtainable with short pulse excitation. The experiments have been carried out[17] in homeotropic samples of the nematic 5CB doped by a small amount of methyl-red (1% by weight) using the second harmonic of a Q-switched Nd:YAG laser (λ = 532 nm), providing pulses of 4ns. The usual set-up for holographic recording has been used under the Mach-Zender configuration: two coherent beams overlap at a small angle on the sample, while the beam coming from a low power He-Ne laser and impinging at normal incidence on it probes the induced grating diffraction.

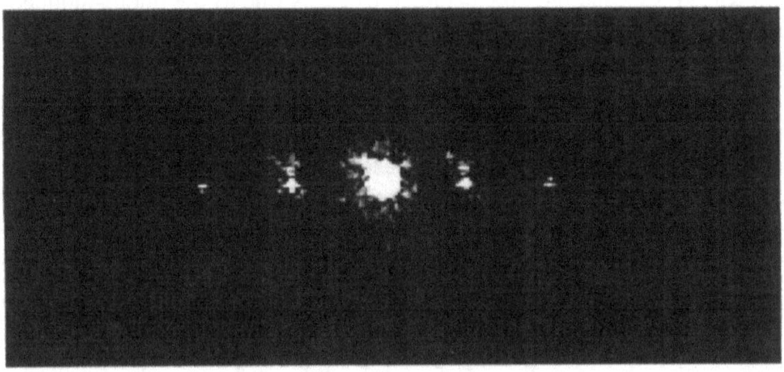

Figure1: Photo of the diffraction pattern of the probe beam after a single 4 ns shot of the exciting beams

In fig.1 is reported the observed dynamic grating excited by a single 4ns pulse of the pump beams, as recorded by a CCD camera. On the first diffraction spot an efficiency of 2% has been measured using a total energy density of the exciting beams lower than 1 mJ/cm^2. The orientational character of the effect has been confirmed also in this case by the light polarization dependence and by the measured grating decay time in the range 25-100 ms. An estimation of the correspondent nonlinear refractive index is obtained in the following way. The measured photo-induced optical anisotropy is $\Delta n = 1.2\ 10^{-3}$ at $I = 2.5\ 10^6\ W/cm^2$, however since the pulse duration is $\tau_p = 4$ ns, while the measured characteristic time of the effect is $\tau_{off} = 0.7$ s, the nonlinear refractive index is calculated as:

$$n_2 = \frac{\Delta n}{I} \cdot \frac{\tau_{off}}{\tau_p}$$

In this way we get $n_2 = 0.8\ cm^2/W$ in agreement with the results obtained by Khoo under c.w. excitation.

The achievement and control of such a huge nonlinear response is obviously of great interest for applications since it may really give rise to a new technology for Spatial Light Modulators for wavefront correction and image processing. An example of such application has been recently given by Khoo and coworker[18]. While the present technology exploits the liquid crystal valve (LCV), namely an artificial nonlinear medium with sensitivity in the range of mW/cm^2 and spatial resolution limited to 30-40 lines/mm, with these materials the same sensitivity can be obtained with much higher spatial resolution (>1000 lines/mm) and with a very simple technological architecture.

References

[1] B. Ya. Zel'dovich, N.F. Pilipetstskii, A.Sukhov and N.Tabiryan, *JETP Lett. 31*, 264 (1980)
[2] I.C.Khoo and S.L.Zhuang, *Appl.Phys.Lett. 37*, 3 (1980)
[3] F.Simoni, Nonlinear Optical Properties of Liquid Crystals and Polymer Dispersed Liquid Crystals, *World Scientific*, Singapore 1997
[4] F.Simoni and R.Bartolino, *Opt.Comm. 53*, 210 (1985)
[5] F.Bloisi, L.Vicari, F.Simoni, G.Cipparrone and C.Umeton, *J. Opt. Soc. Am. B5*, 2462 (1988)
[6] I.C.Khoo, *Phys.Rev. A25*, 1636 (1982)
[7] I.Janossy, A.D.Lloyd and B.S.Wherrett, *Mol.Cryst.Liq.Cryst. 179*, 1 (1990)
[8] I.Janossy and A.D.Lloyd, *Mol.Cryst.Liq.Cryst. 203*, 77 (1991)
[9] I.Janossy, L.Csillag and A.D.Lloyd, *Phys.Rev. A44*, 8410 (1991)
[10] L.Marrucci and D.Paparo, *Phys.Rev. E56*, 1765 (1997)
[11] I.Janossy and Szabados, *Phys.Rev. E58*, 4598 (1998)
[12] E.V.Rudenko and A.V.Sukhov, *JETP Lett. 59*, 142 (1994)
[13] I.C.Khoo, H.Li and Y.Liamg, *Opt.Lett.19*, 1723 (1994)
[14] I.C.Khoo, M.-Y. Shih, M.V.Wood, B.D.Guenther, P.H.Chen, F.Simoni,.S.Slussarenko, O.Francescangeli and Liana Lucchetti, *Proc.of the IEEE 87*, 1897 (1999)
[15] I.C.Khoo, *IEEE J.Quantum Electron.32*, 525 (1996)
[16] I.C.Khoo, S.Slussarenko, B.D.Guenther, M.-Y. Shih, P.Chen and W.V.Wood, *Opt.Lett.23*, 253 (1998)
[17] L.Lucchetti, S.Slussarenko, O.Francescangeli and F.Simoni, submitted to *Appl.Phys.Lett.*
[18] M.-Y. Shih, A.Shishido, P.Chen, W.V.Wood. and I.C.Khoo, *Opt.Lett.23*, 253 (1998)

CHARACTERIZATION OF AlGaAs/AlAs WAVEGUIDES FOR OPTICAL PARAMETRIC GENERATION

G. LEO[1], A. DE ROSSI[2], V. BERGER[2], G. ASSANTO[1]

1) *Dept. of EE & National Institute for the Physics of Matter INFM-RM3 University "Roma Tre", Via della Vasca Navale 84, 00146 Rome-Italy*
2) *Thomson-CSF, Laboratoire Central de Recherches Domaine de Corbeville, 91400 Orsay-France*

Abstract

We report on our research activity towards the realization of an AlGaAs-based integrated optical parametric oscillator, in a configuration exploiting form birefringence for phase-matching. We emphasize the role of a careful characterization in achieving parametric fluorescence and oscillation. In particular, we present the results of two complementary optical techniques, providing a more accurate estimate of refractive indices than extrapolated from existing models: grating-assisted effective index evaluation, and modal birefringence measurement through surface emitted second harmonic generation. We report on our research activity towards the realization of an AlGaAs-based integrated optical parametric oscillator, in a configuration exploiting form birefringence for phase-matching. We emphasize the role of a careful characterization in achieving parametric fluorescence and oscillation. In particular, we present the results of two complementary optical techniques, providing a more accurate estimate of refractive indices than extrapolated from existing models: grating-assisted effective index evaluation, and modal birefringence measurement through surface emitted second harmonic generation.

The state of the art in growth and processing of semiconductor heterostructures grants "refractive-index engineering" great opportunities towards the realization of fully tailored integrated optics structures, in analogy to band-gap engineering in microelectronics and optoelectronics. [1] However, design constraints for resonant or phase-matched geometries are usually more strict than the standard compliance on thickness uniformity in μm-thick MBE-grown multilayers. Moreover, the required accuracies often exceed those available from known refractive indices of most semiconductor compounds: for AlGaAs, existing models of refractive index provide a resolution to the second decimal digit. [2-4] Finally, the optical properties of oxidized AlAs layers, recently employed for both optical confinement and form-birefringence

A. Driessen (ed.), Nonlinear Optics for the Information Society, 35–41.

engineering, [5] strongly depend on layer thickness and oxidation conditions. In particular, for realizing an integrated optical parametric oscillator (OPO), the effective indices of the interacting modes need be known with a precision $\geq 10^{-3}$.
In the following we illustrate two complementary optical-characterization techniques. The first is an optimized version of standard m-line spectroscopy, and was employed at 1310 and 1550 nm to provide effective indices with errors as low as ±0.0005. When crossing such data with independent, nondestructive X-ray measurements of multilayer thicknesses, layer indices could be estimated within ±0.0005.
The second technique, performed at 1319 nm, allows a one-shot measurement of the TE-TM modal birefringences of a nonlinear waveguide through the analysis of surface-emitted second harmonic generation (SESHG).

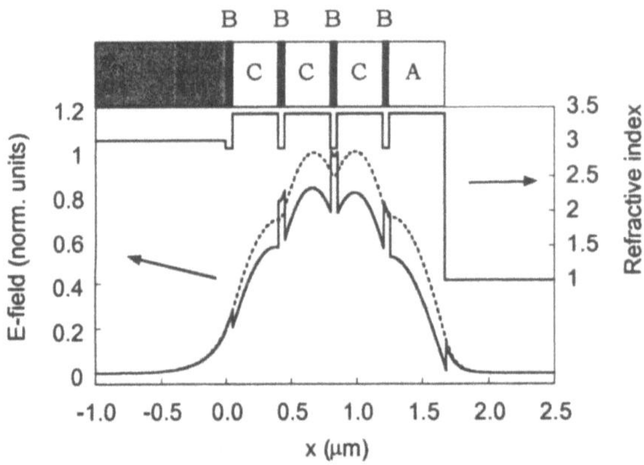

Figure 1: Refractive index profile of the waveguide, and tranverse field profiles at λ=1.31μm for TE_0 (dotted line) and TM_0 (solid line) modes. The multilayer structure is also sketched on the top (GaAs, white; AlAs, black; $Al_{0.75}Ga_{0.25}As$, gray), with the corresponding GaAs and AlAs thicknesses A, B, C.

Measurements of thickness and guided-wave indices

Our samples were grown by MBE on a GaAs wafer with the structure: $Al_{0.75}Ga_{0.25}As$ (1500nm) / AlAs (50nm) / 3 ×[GaAs (350nm) / AlAs (50nm)] / GaAs (420nm), as sketched in Fig. 1. In the figure we also show the transverse electric-field profiles at λ=1.31μm for the TE_0 and TM_0 eigenmodes, as obtained by standard calculations based on Afromowitz model. [2] To ascertain the actual thicknesses of the multilayer, we resorted to X-ray reflectometry at grazing incidence. This is only sensitive to the mean electronic density and therefore, neglecting absorption, "sees" a thin layer as a homogeneous medium. [6] By scanning the angle of incidence, a system of "Kiessig" fringes appears, the period of which is related to layer thickness. [7] Thus, by Fourier analyzing the resulting interference pattern, a remarkable accuracy of ±0.05nm can be achieved for layers of about 50 nm, and the interface separations can be inferred as well. X-ray reflectometry was performed at the $Cu_{K\alpha 1}$ wavelength (λ=0.1540562nm) using a high resolution Seifert PTS goniometer equipped with a front double Ge(220) monochromator and a multilayer mirror to enhance the incident intensity. Back Soller slits and a knife located at 60μm from the sample surface allowed to reduce the

background down to 0.05cps. Thus, a dynamic range of 10^7 (in intensity) can be achieved in reflectivity, as visible in the typical results in Fig.2.
The n_{eff} measurements were performed by distributed input-coupling of radiation to guided modes thru a grating, and by endfire output-coupling onto a large-area Ge photodiode. The experimental setup is sketched Fig.3. The sample was placed on a computer-controlled rotation stage with 0.001° resolution, and the experiment consisted in scanning the angle of incidence θ while synchronously detecting peaks in guided light at the output. A periodic pattern was realized on the waveguide by a standard holographic set-up consisting of a cw He-Cd source at λ=442 nm and a Lloyd-mirror. [8] After exposure and development, a submicron photoresist (Shipley-1805) grating was obtained on the surface. This was either employed as such or subsequently etched in a H_2SO_4 solution. Wavevector conservation in coupling an input field of wavelength λ into a guided mode of effective index n_{eff} through a first-order grating of period Λ implies: $n_{eff} = sin\theta + \lambda/\Lambda$. The measurement yields an angular spectrum related to the n_{eff} value of all the modes supported at a given polarization. Zero angle setting and evaluation of grating period were carried out with a He-Ne source (633 nm), while data

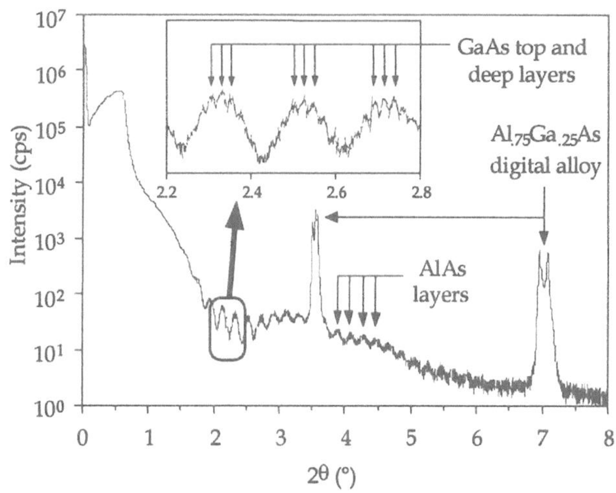

Figure 2: X-ray reflectivity of the waveguide multilayer. Long period oscillations are due to the AlAs layers, while the short-period ones (inset) refer to the thicker GaAs layers.

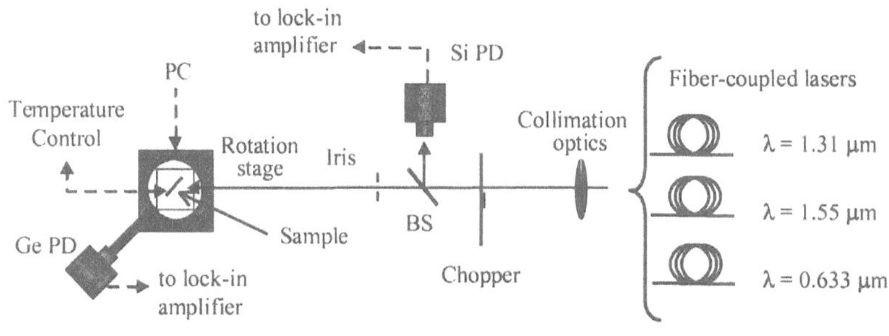

Figure 3: Experimental setup for n_{eff} measurements.

were collected at several wavelengths around λ=1.31 and 1.55 μm. After a Gaussian fit of the discrete resonances, we achieved precisions within 0.005÷0.01°, mainly limited by the stepper-motor repeatability. With temperature stabilization within 0.01°C and λ selected by DFB sources, the error bar in n_{eff} is mainly related to the uncertainty on Λ. A typical angular scan at λ=1.31 μm is shown in Fig. 4 for a planar waveguide with a 0.4 μm-thick photoresist grating and Λ=437.98 ± 0.04 nm. Through proper control of input polarization, the six peaks can be unambiguously identified as the three TE and TM lowest-order modes. The values of n_{eff} were inferred with an absolute precision of $\pm 5 \times 10^{-4}$ (comparable to what reported by Kaufman *et al.* [9]), adequate for phase matching in parametric devices.
From independently measured thicknesses and n_{eff}, we derived the refractive indices of the multilayer at the two wavelengths above, using standard inversion of the eigenmode problem. These values are listed in Table 1, along with predictions from Afromowitz's and Adachi's models, respectively. [2,3] While error bars associated to n(GaAs), n(AlAs), and n(AlGaAs) are somewhat different due to unequal role of each layer on dispersion and n_{eff}, none of the two models is to be consistently preferred. Measured values therefore constitute our most reliable guideline, especially when considering the influence of growth conditions on the optical properties of the multilayer under test.

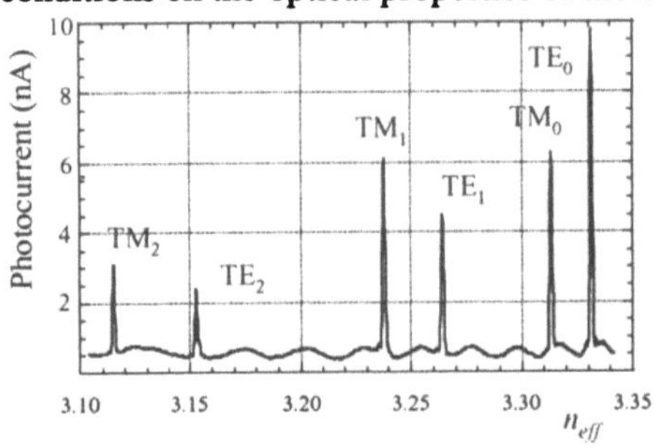

Figure 4: Output signal vs. incidence angle at λ=1.31μm in a planar waveguide.

TABLE 1 Refractive indices of the multilayer films at λ=1.31 and λ=1.55μm: measured and calculated values after Afromowitz [2] and Adachi [3].

	Our Experiment		Afromowitz		Adachi	
	1.31μm	**1.55μm**	**1.31μm**	**1.55μm**	**1.31μm**	**1.55μm**
GaAs	3.4058±.0006	3.3742±.0008	3.411	3.377	3.449	3.431
75% AlGaAs	3.08±.01	3.027±.005	3.026	3.008	3.061	3.045
AlAs	2.924±.004	2.893±.003	2.909	2.894	2.924	2.909

Birefringence evaluation via SESHG

The main concept underlying SESHG [11-13] is sketched in Fig.5a for a nonlinear slab waveguide: the wavevector mismatch $\Delta\beta=\beta^{+}-\beta^{-}$ between two counter-propagating modes at the fundamental frequency ω sets the out-of-plane emission angle θ for the radiated second harmonic 2ω. This angle depends on the difference ΔN between the

n_{eff}'s of the counter-propagating fields at ω. In the nonlinear polarization, the effective nonlinearity d_{eff} depends on both the tensor structure and the polarizations of the fields at ω. In our case, (100) GaAs substrates, $d_{eff}=d_{123}$ with SESHG driven by TE-TM interactions, i.e., $\Delta\beta=\beta^{TE}-\beta^{TM}$. To excite SESHG we excited both TE and TM eigenfields at one input facet of the sample, exploiting Fresnel reflection at the opposite facet to generate the required counter-propagating modal components at ω. In this case (see Fig. 5b), two out-of-plane waves are emitted (interactions E_{TE}^{+}-E_{TM}^{-} and E_{TM}^{+}-E_{TE}^{-}). Being these associated to a stationary wave of the nonlinear polarization $P^{(2\omega)}\propto\cos(\Delta\beta z)=\cos(\omega\Delta N\ z/c)$, it is possible to evaluate the birefringence ΔN by fringe counting in the near-field (NF) distribution $I^{(2\omega)}(z)\propto|P^{(2\omega)}|^2$. [12] In a multimode waveguide, however, each TE-TM pair produces a distinct NF fringe pattern at 2ω. For different launch conditions, the NF at 2ω may appear as either a single fringe system (TE_{00}-TM_{00} interactions), or a superposition of sinusoidal patterns (several mode couplings). In the latter case, the Fourier-transformed intensity distribution $I^{(2\omega)}$ in the far field (FF) can spatially resolve the single SESHG processes, such that one observation can provide complete information on both birefringence and modal content of multimode waveguides.

We performed experiments on AlGaAs waveguides grown by MBE as: $Al_{0.7}Ga_{0.3}As$ (substrate) / AlAs (200nm) / 2×[$Al_{0.1}Ga_{0.9}As$ (270nm) / AlAs (25nm)] / $Al_{0.1}Ga_{0.9}As$ (270nm) / AlAs (62nm) / $Al_{0.7}Ga_{0.3}As$ (300nm) / GaAs (20nm). The planar structures were reactive-ion etched to create 1μm-thick ridges with 3 to 9 μm widths. We end-fire coupled 100ps pulses at λ=1.32 μm from a mode-locked Nd-YAG laser into waveguides a few mm long (quasi-cw interaction), using linearly polarized light at 45° relative to the substrate. The ridges supported the four lowest-order quasi TE and TM modes, and either NF or FF SESHG images at λ =0.66μm were collected by a CCD camera. In the NF (see Fig. 6a), for a sample length L=3.23±0.02mm, by counting m=68.5±0.5 fringes, we obtained $\Delta N=N(TE_{00})-N(TM_{00})=m\lambda/2L=0.0140\pm0.0002$.

A typical result of the FF SESHG imaged onto the CCD with an *f*=2.5cm lens is shown in Fig. 6b, where each pair of δ functions (Fourier- transform of a sinusoid) refers to the corresponding set of counterpropagating modes. [13] From Fig. 6b it is straighforward to infer the ΔN's, as presented in Table 1 along with the case $N(TE_{10})-N(TM_{10})$, not visible in the photograph but readily obtained by slightly changing the launch conditions. The errors depend on the uncertainties associated to both *f* and pixel discretization, the latter even for the near field. In Table 2 the calculated ΔN's and the related uncertainties are also listed, as derived from Afromowitz dispersion model [2]. As shown in Table 2, this simple single-shot technique allows the accurate estimate of birefringence between all the existing modes at ω, in excellent agreement with the model. Finally, we employed SESHG to investigate waveguides with a higher birefringence, i.e., AlGaAs/AlAs samples as described above but after lateral oxidation of the AlAs layers. As expected, two features are apparent in Fig. 6c: on the one hand the ΔN between TE_{00} and TM_{00} is greatly enhanced by oxidation, i. e., by lowering the index of the inserted thin layers (AlAs→AlOx) [5] from 0.0138 to 0.079. On the other hand $N(TE_{10}) \ll N(TM_{00})$ in the non-oxidized samples, while oxidation lowers $N(TM_{00})$ far more than $N(TE_{10})$, thus explaining the related reduction in ΔN from 0.125 to 0.110.

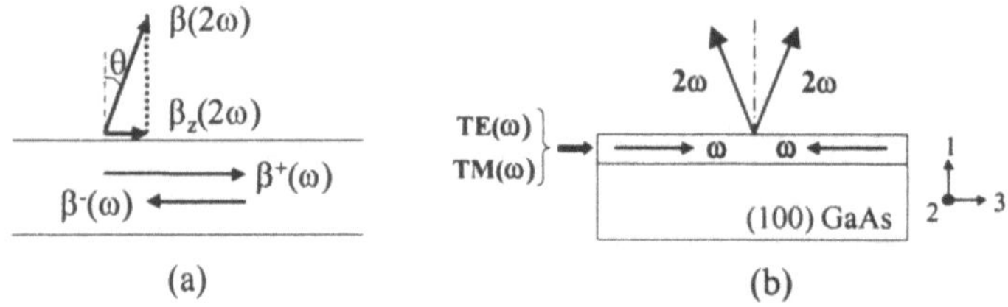

Figure 5: (a) Phase-matching scheme of SESHG at 2ω from two interacting guided waves at ω. (b) Sketch of SESHG with two out-of-plane emissions at opposite angles.

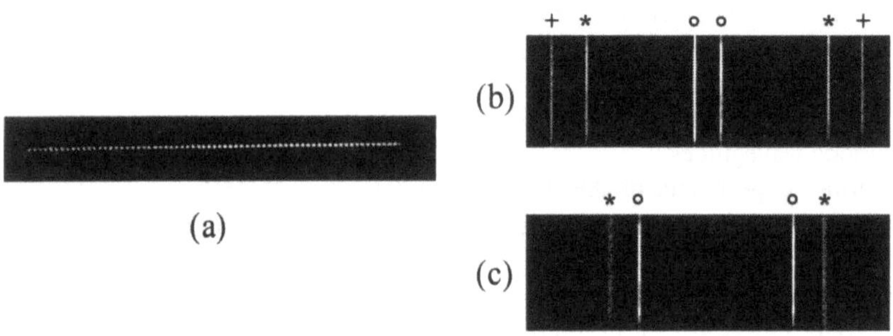

Figure 6: (a) Typical near-field intensity distribution $I^{(2\omega)}$. (b) Far-field picture of SESHG modes from an AlGaAs ridge 7 μm wide, for TE_{00}-TM_{00} (°), TE_{10}-TM_{00} (*), TE_{00}-TM_{10} (+). (c) Same as (b), after selective oxidation of AlAs layers.

TABLE 2: Measured and calculated TE-TM birefringences for a non-oxidized ridge.

interacting modes	*measured ΔN*	*calculated ΔN*
TE_{00} - TM_{00}	0.0138±0.0008	0.014±0.001
TE_{10} - TM_{00}	0.125±0.002	0.122±0.002
TE_{00} - TM_{10}	0.160±0.002	0.158±0.002
TE_{10} - TM_{10}	0.0220±0.0008	0.022±0.001

Conclusion

In order to achieve an accuarate enough knowledge of the optical parameter of novel GaAs-based nonlinear guiding structures, we have carried out a twofold approach. By combining accurate effective-index measurements to X-ray reflectometry at grazing incidence, we achieved state-of-the-art figures for both refractive/effective indices and multilayer thicknesses. [10] Moreover, we demonstrated that SESHG can be effectively used to simultaneously and accurately determine all modal birefringences of parametric waveguides. These techniques are a valuable tool towards realizing a GaAs-based integrated OPO.

This research was financed by the EU, under ESPRIT Project OFCORSE II.

References

[1] See, for example, V. Berger, Photonic Crystals for Nonlinear Optical Frequency Conversion, *Confined Photon Systems*, H. Benisty, J. Gérard, R. Houdré, J. Rarity and C. Weisbuch Eds. (Springer, Berlin, 1999).

[2] M. A. Afromowitz, *Solid State Commun., vol.15*, p.39 (1974).

[3] S. Adachi, *J. Appl. Phys., vol.58*, p.R1 (1985).

[4] E. D. Palik, *Handbook of optical constants of solids* (Academic, San Diego, 1985).

[5] A. Fiore, V. Berger, E. Rosencher, N. Laurent, S. Theilmann, N. Vodjani, J. Nagle, *Appl. Phys. Lett., vol.68*, p.1320 (1996).

[6] B. Vidal and P. Vincent, *Appl. Opt., vol.23*, p.1794 (1984).

[7] H. Kiessig, *Ann. Physik, vol.10*, p.769 (1931).

[8] X. Mai, R. Moshrefzadeh, U. J. Gibson, G. I. Stegeman, and C. T. Seaton, *Appl. Opt., vol.24*, p.3155 (1985).

[9] R. G. Kaufman, G. R. Hulse, K. A. Stair, T. E. Bird, G. P. Devane, and A. L. Moretti, J. *Appl. Phys., vol.77*, p.1747 (1995).

[10] G. Leo, C. Caldarella, G. Masini, A. De Rossi, G. Assanto, O. Durand, M. Calligaro, X. Marcadet, V. Berger, "X-ray and optical characterization of multilayer AlGaAs waveguides", *Appl. Phys. Lett., in press* (2000).

[11] P. J. Vella, R. Normandin, G.I. Stegeman, *Appl. Phys. Lett., vol.38*, p.760 (1981).

[12] D. Vakhshoori, M. C. Wu, S. Wang, Appl. *Phys. Lett., vol.52*, p.422 (1988).

[13] A. Fiore, V. Berger, E. Rosencher, S. Crouzy, N. Laurent, J. Nagle, *Appl. Phys. Lett., vol.71*, p.2587 (1997).

NONLINEAR OPTICS FOR QUANTUM COMMUNICATION

H. ZBINDEN
Université de Genève, GAP-O 20
Rue de l'Ecole-de-Médecine, CH-1211 Genève 4

Abstract

Presently, Quantum Communication is a very hot topic. People are dreaming about quantum computers and teleportation. However, the experimental problems are huge if not insurmountable and quantum cryptography is the only practical application that emerges so far.
We give a brief introduction in quantum cryptography, quantum computing and teleportation. In many experiments nonlinear optics is involved in a way or another. We discuss what kind of developments in nonlinear optics would be fruitful for experiments related to Quantum Communication.

Full article not available

A. Driessen (ed.), Nonlinear Optics for the Information Society, 43.

ALL-OPTICAL DIODE IN A QUASI-PHASE-MATCHED $LiNbO_3$ WAVEGUIDE

K. R. PARAMESWARAN, M. M. FEJER[1], K. GALLO[2], G. ASSANTO[3]

1) E. L. Ginzton Laboratory, Stanford University, Stanford, CA 94305-4085 USA

2) present address : Laboratoire de Physique de la Matière Condensée – Université de Nice, Parc Valrose – 06108 Nice – France

3) National Institute for the Physics of Matter & Department of Electronic Engineering Terza University of Rome, Via della Vasca Navale 84 - 00146 Rome – Italy

Keywords

Nonlinear Optics, Quasi-Phase-Matching, Lithium Niobate, Waveguide.

Abstract

We report on the first demonstration of an integrated power-dependent all-optical isolator at 1.55 μm based on proper tailoring of a Quasi-Phase-Matching grating in an Annealed Proton-Exchanged $LiNbO_3$ channel waveguide. For input powers beyond 1.5 W we observe the onset of a noticeable non-reciprocal response, in good agreement with predictions, reaching a contrast as high as 90% with an input power of 3.1W.

1. Introduction

In analogy to its electronic counterpart, an all-optical "diode" (AOD) is a device allowing unidirectional propagation of a signal at a given (optical) wavelength (λ). Ideally, its transmission would equal unity in the "forward" direction ($T_\lambda^+=1$) and zero in the opposite one ("backward":$T_\lambda^-=0$), yielding a unitary contrast C = $T_\lambda^+(T_\lambda^+-T_\lambda^-)/(T_\lambda^++T_\lambda^-)$.

Recent years witnessed a renewed interest in integrated-optic alternatives to standard bulk Faraday rotators to perform optical isolation in fiber telecommunication systems. However the thin film devices proposed to date still rely on magneto-optics [1][2]. Here we undertake a completely different approach to realise an integrated-optic power-dependent AOD, relying on non-reciprocal second harmonic generation (SHG) [3][4], and demonstrate its operation at 1.55μm in $LiNbO_3$. The device exploits the mature technology of Quasi-Phase-Matching (QPM) grating and waveguide fabrication in

A. Driessen (ed.), Nonlinear Optics for the Information Society, 45–49.

$LiNbO_3$ by means of electric-field periodic poling [5] and annealed proton-exchange (APE) [6], respectively.
The spatially non-reciprocal response stems from the combination of the quadratic nonlinearity of the $LiNbO_3$ medium with a built-in asymmetry along the propagation direction of light. As opposed to a previously reported experiment, where the *linear* properties were asymmetrically altered by depositing a cladding layer over half of the waveguide [4], we engineer the response of our AOD by tailoring the *nonlinear* QPM grating profile. This proves to be a more flexible and reliable way of controlling the device response and maximising its contrast.
In good agreement with the theoretically predicted behaviour, we have demonstrated, for the first time, an optical diode at 1.55 μm exhibiting an isolation figure of merit close to the ideal value of 100%.

2. Design

The AOD is based on the insertion of an aperiodic domain inside an otherwise perfectly uniform QPM grating as sketched in Fig. 1. The structure allows cascading of two SHG interactions (occurring in the uniform sections L1 and L2) while locally changing (at z = L1) the phase relationship between the fundamental and the second harmonic waves by an amount which is determined by the length of the aperiodic domain.

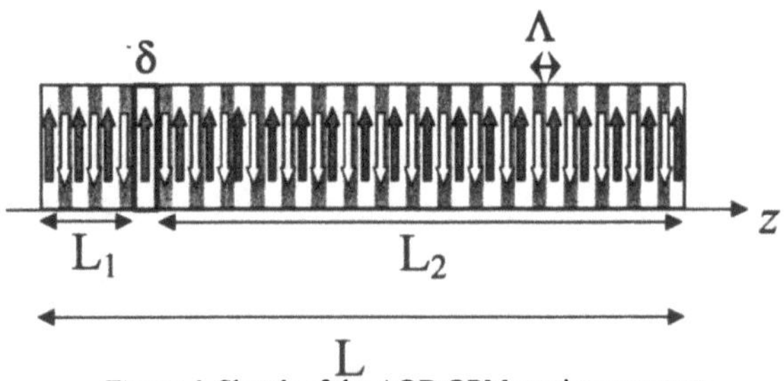

Figure 1. Sketch of the AOD QPM grating structure.

The nonlinear grating period (Λ) is equal to twice the coherence length of the SHG interaction involving the APE waveguide fundamental transverse (TM_{00}) modes at λ and λ/2, while the optimum values for the other parameters (L_1, δ) are determined in a preliminary design stage aimed at maximising the device contrast.

Figure 2 shows a contour plot of the AOD contrast $(T_\lambda^+ - T_\lambda^-)/(T_\lambda^+ + T_\lambda^-)$ as a function of the dephasing $\Delta\theta = 2\pi\delta/\Lambda - \pi$ (induced by the aperiodic domain) and normalised input power $\Gamma^2 = \eta P_{in} L^2$ [η, P_{in} and L being the normalised SHG efficiency ($W^{-1}cm^{-2}$), input power (W) and overall grating length (cm), respectively], calculated for $L_1 = 0.2 \cdot L$. The contours identify contrast values from 20% to 99.9%.
It is apparent that a proper choice of the dephasing allows (for normalised input power ranges beyond Γ^2=10) contrasts in excess of 95% to be reached [7].

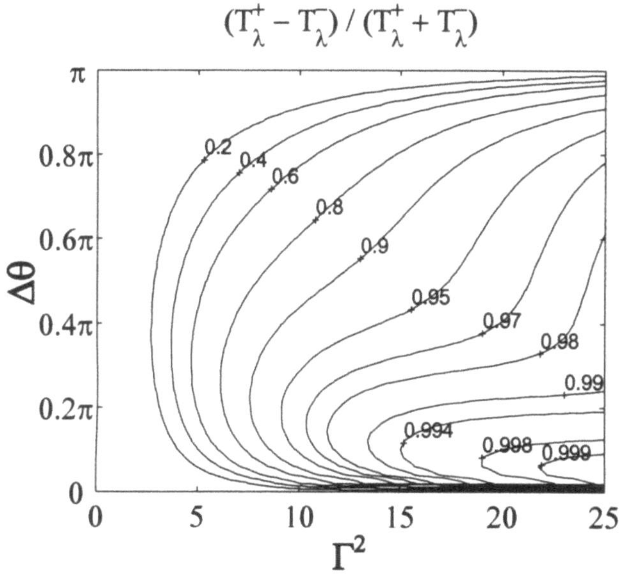

Figure 2. Contour plot of the AOD figure of merit $(T_\lambda^+ - T_\lambda^-)/(T_\lambda^+ + T_\lambda^-)$ versus $\Gamma^2=\eta P_{in} L^2$ and $\Delta\theta=2\pi\delta/\Lambda-\pi$ for $L_1=0.2\cdot L$. The labels identify the contrast values on the contours.

The predicted behaviour can be explained in terms of *quadratic cascading* effects [8]. In the "forward" direction, for input powers beyond a certain threshold, cascaded up- and down- conversion processes can take place in the L_1 and L_2 sections, respectively. In the structure of Fig.1 the standard SHG evolution beginning in the first segment (L_1) is perturbed by the dephasing $\Delta\theta$ which triggers a *seeded* SHG [9] starting at $x = L_1+\delta$. A proper choice of the value of $\Delta\theta$ makes the *seeded* SHG result in a recovery of the power at 1.55 μm at the output (by down-conversion of the second harmonic).
At the same input power level in the "backward" direction, the first standard SHG taking place in the longer segment (L_2) can result in complete depletion of the fundamental before reaching the defect δ, so that the backward transmission at λ is zero regardless of the value of $\Delta\theta$.

3. Device realisation and characterisation

After identifying the QPM grating profiles most suitable for implementing an AOD, we 'wrote' them by electric field poling on a 500 μm thick Z-cut $LiNbO_3$ wafer. The QPM gratings had an overall length L = 4 cm and a period Λ = 14.7 μm. On the diced samples we then fabricated channel waveguides by proton exchange through 12 μm wide mask openings (depth = 0.71 μm) and annealing (26 h at 327.5 °C).
The normalised SHG efficiency η [$W^{-1}cm^{-2}$] and propagation losses α [$dB\cdot cm^{-1}$] were assessed through a low-power CW measurement.
The devices were then tested at high powers by means of a tuneable, externally modulated, external cavity diode laser delivering 20 ns flat-top pulses at 1.55 μm which

were then amplified by an all-fiber system to reach powers of several Watts at the waveguide input (Fig.3).

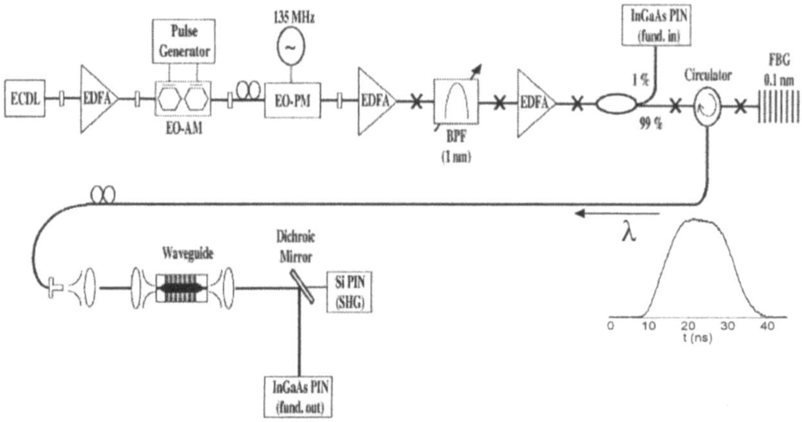

Figure 3. Setup for the characterisation of the AOD at high powers. The inset shows the actual pulse shape at the input of the device. ECDL = External Cavity Diode Laser, EDFA = Erbium-Doped Fiber Amplifier, EO-AM = Electro-Optic Amplitude Modulator, EO-PM = Electro-Optic Phase Modulator (used only for powers exceeding ~5W), BPF = Band Pass Filter, PIN = p-i-n junction photodiode, FBG = Fiber Bragg Grating.

For each peak power P_{in} we recorded the generated SH and the transmitted FF on forward propagation and repeated the measurement for backward propagation (after rotating the sample by 180°).

Fig.4 shows the experimental results (symbols) for a QPM grating with $\delta = 8.1$ μm and $L_1 = 8$ mm in a waveguide with $\eta = 23.6\ W^{-1}\,cm^{-2}$, $\alpha_\lambda = 0.68$ dB/cm, $\alpha_{\lambda/2} = 0.8$ dB/cm, along with the predicted behaviour (solid curves). The reported data correspond to the transmission at 1.55 μm normalised to the value for small P_{in} on forward and backward propagation.

Theory and experiment are in good agreement. As expected, for increasing input powers P_{in} the forward and backward transmissions begin to separate: the *seeded* SHG induced by the defect, makes T_λ^+ increase again after $P_{in} \approx 1.6$ W, whereas T_λ^- does not differ from that of an efficient frequency doubler.

For $P_{in} = 3.1$ W, we obtained a record contrast $(T_\lambda^+ - T_\lambda^-)/(T_\lambda^+ + T_\lambda^-) = 90\%$. We are working at present in order to explore the AOD response at higher excitations where we expect to reach even higher values of the figure of merit (close to unity).

In conclusion, we have designed and demonstrated a power-dependent all-optical diode in a QPM APE waveguide in $LiNbO_3$, operating in the third fiber telecommunications window. The device also confirms the potential of engineering cascaded effects in $\chi^{(2)}$ interactions to realise fully integrated devices for all-optical signal processing.

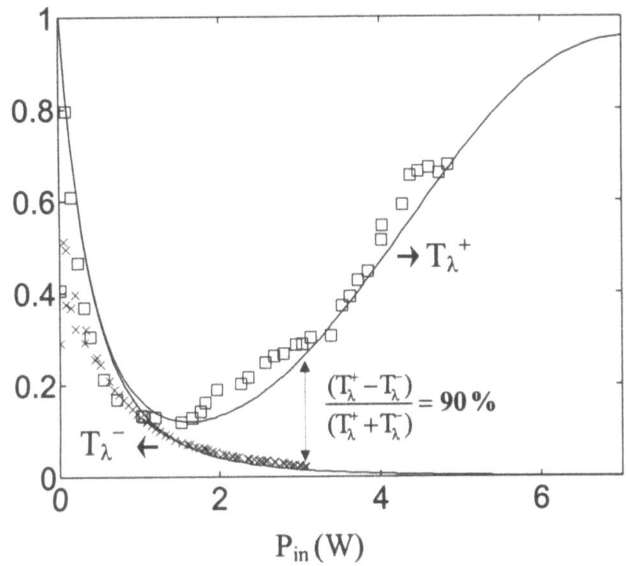

Figure 4. Measured forward (T_λ^+: squares) and backward (T_λ^-: crosses) transmissions normalised to their value for $P_{in} \rightarrow 0$ versus input power P_{in}, along with the theoretical predictions (solid lines).

References

[1] J. Fujita, M. Levy, R. M. Osgood, Jr., L. Wilkens, and H. Dötsch, Waveguide optical isolator based on Mach–Zehnder interferometer, *Appl. Phys. Lett. 76*, 2158–2160 (2000)

[2] T. Mizumoto, H. Yokoi, N. Futakuchi, and Y. Nakano, Optical isolator using a nonreciprocal phase shift with a semiconductor guiding layer", *Proc. LEOS 2000 Annual Meeting, IEEE cat.# 00CH37080*, p.700 (2000)

[3] Yu. P. Svirko and N. I. Zheludev, Reversality of optical interactions in noncentrosymmetric media, *Opt. Lett. 20*, 1809–1811 (1995)

[4] C. G. Treviño-Palacios, G. I. Stegeman, and P. Baldi, Spatial nonreciprocity in waveguide second-order processes, *Opt. Lett. 21*, 1442–1444 (1996)

[5] L. E. Myers, R. C. Eckardt, M. M. Fejer, R. L. Byer, W. R. Bosenberg, and J.W. Pierce, Quasi-phase-matched optical parametric oscillators in bulk periodically poled $LiNbO_3$, *J. Opt. Soc. Am. B 12*, 2102–2116 (1995)

[6] K. R. Parameswaran, M. Fujimura, M. H. Chou, and M. M. Fejer, Low-power all-optical gate based on sum frequency mixing in APE waveguides in PPLN, *IEEE Photon. Technol. Lett. 12*, 654–656 (2000)

[7] K. Gallo and G. Assanto, All-optical diode based on second harmonic generation in an asymmetric waveguide, *J. Opt. Soc. Am. B 16*, 267–269 (1999)

[8] G. Assanto, Quadratic cascading: effects and applications, *Beam Shaping and control with Nonlinear Optics*, F. Kajzar and R. Reinisch eds, *Plenum Press B 369*, 341–374, New York (1997)

[9] G. Assanto, G.I. Stegeman, M. Sheik-Bahae and E. Van Stryland, Coherent interactions for all-optical signal processing via quadratic nonlinearities, *IEEE J. Quantum Electron. 31*, 673–681 (1995)

FEASIBILITY OF ALL-OPTICAL COMPUTATION BY SECOND HARMONIC GENERATED HOLOGRAMS

A. ANDREONI[1], M. BONDANI[1], M.A.C. POTENZA[1], E. PUDDU[2], Y. N. DENISYUK[3]

1) *Dipartimento di Scienze CC, FF e MM, Universita' degli Studi dell'Insubria, and Istituto Nazionale di Fisica della Materia, I.N.F.M., Via Valleggio, 11, 22100 Como, Italy*
2) *Second Faculty of Sciences, Universita' degli Studi dell'Insubria, Via Valleggio, 11, 22100 Como, Italy*
3) *Ioffe Physico-Technical Institute, Politechnicheskaya Street, 26, 194021 St. Petersburg, Russia*

Keywords

holography, optical computing, boolean algebra

Abstract

All-optical computation is shown to be working when boolean algebra operations are performed on optical bits encoded as amplitude modulations in two wavefronts that are made interact in a non-linear crystal endowed with second order non-linearity (β-BaB_2O_4). The sum-frequency interaction reconstructs a holographic image containing the outputs of the desired operations, and difference frequency also demonstrated to be effective. The experiments demonstrate the feasibility of a parallel half-adder and confirm the possibility to work with high density of data, since a nearly diffraction limited resolution can be easily achieved in the holographic image.

Holographic methods of wavefront reconstruction and transformation started to be discovered more than 50 years ago [1], but only after the 70's fast processing of the wavefronts became a property of holography [2]. In recent years these applications stimulated the research on materials useful to increase the degree of parallelism in manipulating optical data [3]. Here we present an application to optical computing of a new dynamic holographic technique endowed with both high spatial and temporal bandwidths.

This method is based on the second-harmonic cross-correlation (SHCC) in a non-linear crystal of the wavefront to be transformed with a wavefront at the same frequency that is not modified during the interaction (reference wavefront). Since our method

A. Driessen (ed.), Nonlinear Optics for the Information Society, 51–54.

transforms optical wavefields in the same way as a suitably reconstructed conventional hologram would do [4][5][6], we call the output "second-harmonic generated (SHG) hologram".
Here we demonstrate the feasibility of binary gates performing boolean algebra operations such as AND and XOR in a suitable optical set-up producing SHG holograms. The output of a Q-switched amplified Nd:YAG laser was split into two beams, *o* (object) and *r* (reference) (Fig. 1). The two inputs *A* and *B* are encoded as amplitude modulation in *o* and *r* by inserting a mask into each beam path: opaque and transparent pixels of the mask represented the LOW and HIGH values. Beam *o* was passed through a diffuser (S) so that the entire crystal received light from all points of the object mask, while *r* was spatially filtered to obtain a plane wave.

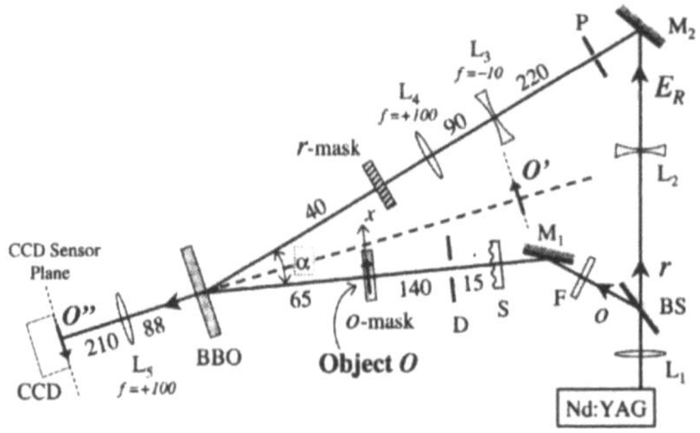

Figure 1

The SHG hologram of the object mask produced by such a system exhibits the amplitude modulation corresponding to the output of an AND gate operating on the inputs.
The implementation of an OR gate is even more straightforward because its output is represented by the simple spatial superimposition of the optical input data. On the contrary, the NOT gate is not obvious. We circumvented this difficulty by using the high space resolution in the SHG images obtained with our method [4][5].
As a rule, together with inputs *A* and *B*, we decided to encode also the complements of inputs *A* and *B* ($\bar{A}$ and $\bar{B}$). The corresponding object and reference masks are shown in Fig. 2. Thus we adapted our set-up to obtain four SHG holograms encoding all possible AND outputs ($A{\cdot}B$, $\bar{A}\cdot B$, $A\cdot\bar{B}$, and $\bar{A}\cdot\bar{B}$) in parallel.

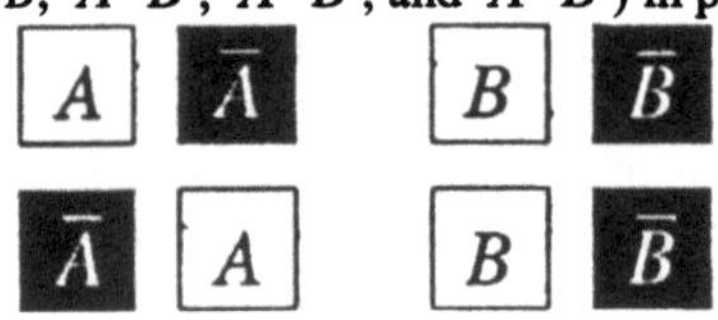

Figure 2

Here we mention the key points of the design of the system. The mask modulating the *r*-beam simply projects a shadow onto the crystal surface. In order to prevent diffraction, this mask contained only four pixels (*i.e.* a single one-digit binary number). Since diffraction does not bring about similar limitations to the object wavefront, we must

simply ensure that inputs A and $\overline{A}$ undergo SHCC in the desired parts of the crystal cross-section. To this aim a diaphragm (D) was inserted after the diffuser S to limit the divergence of the object-illuminating light so that each point-like source at the object plane illuminated about a quarter of the whole crystal. Thus the crystal generated four holograms encoding $A \cdot B$, $\overline{A} \cdot \overline{B}$, $\overline{A} \cdot B$, $A \cdot \overline{B}$, and the superimposition of the last two holograms provided the output of a XOR gate. The holograms thus contain all the outputs needed to realise a half adder, since this device requires an AND gate to provide the carry and a XOR gate for the less significant bit. Results are shown for the case 1+1 (Fig. 3(a)) and 0+1 (Fig. 3(b)).
Preliminary experiments showed that by introducing in the same system a matrix of 16 single-digit binary numbers, A_{ij} (and complements $\overline{A}_{ij}$), it was straightforward to obtain a parallel half-adder. Moreover the parallel operation of the same system was checked for pixels as small as $40 \times 40\ \mu m^2$ encoded in the object beam. In this experiment as the object mask O we used a stainless steel mesh made by 40-μm diameter wire, 40μm spaced. Since light beyond the mesh was diffracted at large angles and invaded the entire crystal surface, we could remove diffuser S and diaphragm D. We obtained the SHG hologram shown in Fig. 3(c).

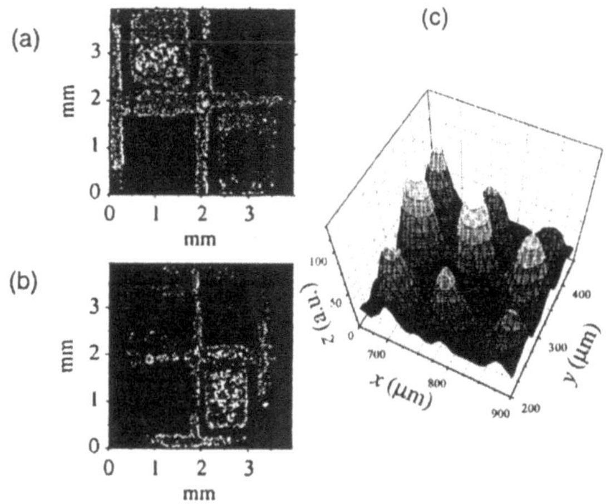

Figure 3

The optical resolution of the SHG hologram of the mesh is not far from the resolving power of our optical system, that is ~44 lines/mm: this is less than twice the spatial frequency of our mesh.
The last point to be discussed concerns the fact that the optical outputs are not at the same frequency as the inputs. We recently showed that, if the object field at ω interacts in a second-order non-linear medium with a reference field at 2ω, the difference-frequency field is a conjugated wavefront reconstructing a hologram (DFG hologram) [7].
Good candidates for the role of object masks seem to be the last-generation liquid-crystal modulators as well as multi-quantum-well modulators. Point-like sources could also be generated by using arrays of vertical-cavity surface-emitting lasers (VCSEL's),

while a diode laser at the same frequency could provide input *B*. The technologies under development for free-space optical interconnections will most likely produce devices useful also to implement a compact half adder based on SHG holograms.

References

[1] D. Gabor, Proc. R. Soc. London, *Ser. A 197*, 454 (1949).
[2] P.Woerdman, *Opt. Commun. 2*, 212 (1970)
[3] Miniewicz, et al, J. Inclus. Phen. Macrocycl. *Chem. 35*, 317 (1999)
[4] Andreoni, et al, *J. Opt. Soc. Am. B 17*, 977 (2000)
[5] Yu. N. Denisyuk, et al, *Opt. Lett. 25*, 890 (2000)
[6] Yu. N. Denisyuk, et al, *Optica i Spectroscopiia* (2000) in press
[7] Andreoni, et al, Phys. *Rev. Lett.*, (2000). Submitted

ALL-OPTICAL WAVELENGTH CONVERSION BY FOUR WAVE MIXING IN A POLYDIACETYLENE WAVEGUIDE

D. GRANDO[1], L. PALCHETTI[1], S. SOTTINI[1], A. FELTRIN[2], G.P. BANFI[2]

1) *Istituto di Ricerca sulle Onde Elettromagnetiche "Nello Carrara", CNR, Via Panciatichi 64, 50127 FIRENZE, Italy Tel. +39 055 4235254, Fax. +39 055 410893E-mail: Sottini@iroe.fi.cnr.it*
2) *Istituto Nazionale di Fisica della Materia and Dipartimento di Elettronica, Università di Pavia, Via Abbiategrasso 209, 27100 PAVIA, Italy Tel. +39 0382 595211; Fax +39 0382 422583 E-mail: Banfi@ele.unipv.it*

Abstract

Wavelength conversion is an important part of the manipulation of optical signals in communication networks. Four Wave Mixing (FWM) is one of the possible mechanisms to achieve all-optical wavelength conversion. In a FWM experiment, starting from a signal pulse at frequency *s* and a pump pulse at frequency *p*, the coherent generation of a beam at frequency $f=2p-s$ and the simultaneous amplification of the pulse at *s* is obtained at the expense of the pump energy via the third order susceptibility of the conversion medium.

Here we propose an all-optical wavelength converter based on FWM in a hybrid conjugated polymer/glass waveguide consisting in a planar graded index ion-exchanged waveguide overlaid by a poly-3BCMU polydiacetylene film. The advantage of this approach is found in the elevated non resonant nonlinearity of conjugated polymers leading to efficient, fast, and large bandwidth wavelength conversion.

We demonstrated frequency conversion in the ps regime. The wavelength of the signal, pump and newly generated beam were equal to 1.286 μm, 1.266 μm and 1.246 μm, respectively. With an effective interaction length of 4.5 mm and an estimated non resonant susceptibility of $1.3\times10^{-19}\ m^2/V^2$ we observed a conversion efficiency of 1% at $100 MW/cm^2$. The discrimination of the different frequencies was achieved by using a grating integrated in the waveguide. The device seems to be a valuable step towards the realisation of a fully integrated wavelength converter.

Full article not available

A. Driessen (ed.), Nonlinear Optics for the Information Society, 55.

ALL-OPTICAL SWITCHING DEVICES IN INTEGRATED OPTICS CYLINDRICAL MICRORESONATORS

An experimental feasibility study

D.J.W. KLUNDER, J. KREMER, F.S. TAN, G. SENGO, E. KRIOUKOV *), C. OTTO*), H.J.W.M HOEKSTRA, A. DRIESSEN
*Lightwave Devices Group, Biophysical Techniques Group *), MESA+ Research Institute, Faculty of applied physics, University of Twente*
PO BOX 217, 7500 AE Enschede, The Netherlands, Phone: +31-53-489 2743, Fax: +31-53-489 3343, email: d.j.w.klunder@tn.utwente.nl.

Key words

All-optical switching, integrated optics, cylindrical microresonators.

Abstract

The feasibility of cylindrical microresonators for all-optical switching applications is discussed both theoretically and experimentally. Using experimental results of high finesse cylindrical microresonators we give estimates for the switching powers.

1. Introduction

Cylindrical microresonators (MRs) have been proposed as all-optical switch [1], microlasers [2], add-drop filters [3], [4] and they open the route to large area reduction of complex integrated optics devices [5]. They have been studied both theoretically (e.g., [3]) and experimentally (e.g., [4], [6], [7], [8]). In this contribution, we will discuss the feasibility of cylindrical microresonators for all-optical switching applications both theoretically and experimentally. First the basic principles of cylindrical microresonators and the key parameters for all-optical switching are discussed. Next we discuss experimental results of high finesse cylindrical microresonators and discuss the most promising microresonator configuration for all-optical switching. Based on this configuration, we will propose an optimised configuration for microresonator based all-optical switches and give estimates for the powers required for switching.

A. Driessen (ed.), Nonlinear Optics for the Information Society, 57–62.

2. Cylindrical Microresonators and all-optical switching

A top view of a waveguide-coupled cylindrical microresonator (MR) is given in Fig. 1.

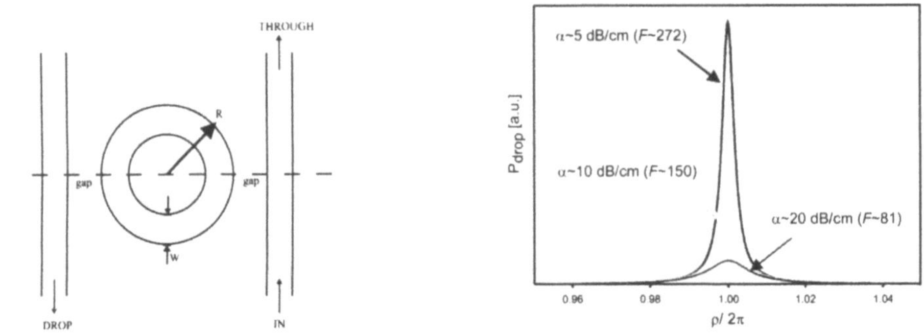

Figure 1. Top view of a symmetrically waveguide-coupled cylindrical microresonator (left), Power in the drop port as a function of the roundtrip phaseshift, ρ and the propagation losses (α) as parameter (right).

A small fraction (κ) of the field of the input straight waveguide (SW) is coupled evanescently with the MR mode. While propagating inside the MR, the modal amplitude of the MR mode decreases due to the propagation losses (α) inside the MR. After half a roundtrip once again a small fraction (κ) of the MR mode is dropped. The field remaining in the MR completes the roundtrip and a small fraction (κ) of the modal amplitude is coupled into the through port and interferes with the fraction of the input field propagating directly to the through port. The field remaining in the MR interferes with the field just coupled in. For a MR on-resonance, a coherent build-up of the intra-cavity power (P_{cav}) results and a large fraction of the input power is dropped (P_{drop}). Important parameters are the free spectral range, FSR (i.e., distance in wavelengths between the resonance peaks) and the finesse F=FSR/FWHM, with FWHM as the full width at half maximum of P_{drop}.

As an illustrative example, we have plotted P_{drop} as a function of the roundtrip phaseshift ρ in Figure 1. It can be seen that P_{drop} can be switched on/off by a small change (i.e. $\Delta\rho/2\pi{\sim}0.02$ for α~5 dB/cm) in the roundtrip phaseshift. The roundtrip phaseshift can be changed by either changing the wavelength or the effective index of the MR mode. As an alternative, assume a MR on resonance (i.e., ρ=0). The power in the drop port (P_{drop}), see also Figure 1, decreases while increasing the propagation losses. Besides switching the MR on/off by changing the roundtrip phaseshift, we can also switch the MR by changing the propagation losses.

For a MR with a Kerr-type non-linearity, both the roundtrip phaseshift (i.e., via the intensity dependent refractive index (IDRI)) and the propagation losses (e.g., via two photon absorption (TPA)) are proportional to the intra-cavity intensity and the overlap between the non-linear medium and the MR mode. In order to minimize the switching power, we have to optimise the ratio between intra-cavity intensity and the input power:

$$\frac{\overline{I}_{cav}}{P_{in}} = \frac{G}{S_{eff}} \qquad (1)$$

with: $G=P_{cav}/P_{in}$ and S_{eff} the effective area of the MR-mode.

The overlap between the non-linear material and the MR-mode can be optimised by, e.g. using the non-linear material as core.

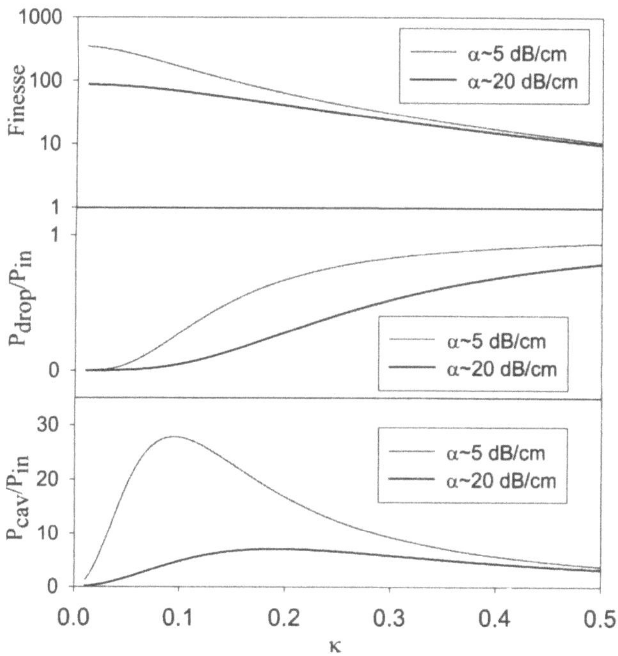

Figure 2. Optimization of a cylindrical microresonator (radius R~25 μm) by varying the coupling factor (κ)

Assuming a given α, the MR can be optimised by varying κ, see Figure 2. The maximum enhancement of intra-cavity power G_{max} decreases monotonously and requires stronger coupling with the SW mode in order to compensate for the losses inside the ring for increasing α: G_{max}=27.8 and 7.0 for α=5 and 20 dB/cm respectively. In this paper we will limit the discussion to polymer based non-linear materials, with relatively large Kerr-type non-linearities (n_{2I}~10^{-17}-10^{-14} m^2W^{-1}) and fast response times (τ<1 ps) due to the pure electronic origin of the non-linearities (see e.g., [9]). Assuming a polymer core layer (i.e., almost perfect overlap between the non-linear material and the MR-mode), an effective area S_{eff}~0.25 μm^2, wavelength λ=1 μm, radius R=25 μm and α=5 dB/cm (i.e., G_{max}=27.8 and *F*~272) this results in a FWHM switching power at the input port of: P_{FWHM}~130 mW.

3. Experimental Results

One way to classify MRs is by distinguishing between laterally (Figure 3 (left)) and vertically (Figure 3 (right)) waveguide-coupled MRs.

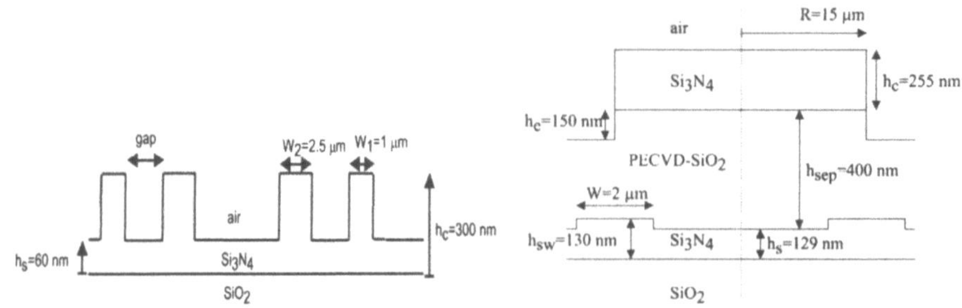

Figure 3. Schematic cross-section of a laterally coupled microring resonator with a radius of R=25 µm (left) and vertically coupled microdisk resonator with a radius of R=15 µm (right).

Laterally waveguide-coupled MRs allow for simple single-mask processing but require an accurate control of the gap between the MR and the SW (which determines the coupling factor, κ).

First consider the microring resonator realized in SiON-technology [10] with air cladding, see also Figure 4 (left). From a detailed fitting procedure we obtained an 'experimental' finesse, F~15, 136 and propagation losses α~11.2 dB/cm for gaps of 500, 750 nm. From the fitting procedure we could also extract κ and calculate the corresponding intra-cavity power enhancement G. Based on the 'experimental' α we found a maximum G of G_{max}=12.5 corresponding to a gap of 500<gap<750 nm.

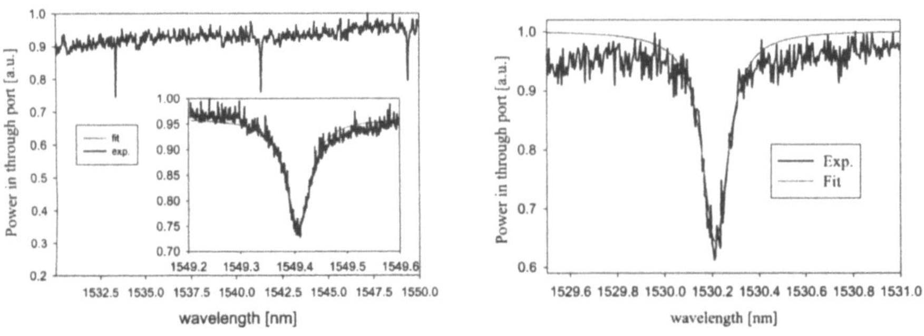

Figure 4. Measured and fitted power in the through port of a laterally coupled microring resonator (radius R=25 µm, gap=750 nm) with air (left) and PMMA (right) cladding layer.

We investigated the effect of a cladding layer (e.g., a polymer with Kerr-type non-linearity) on top of the microring both theoretically and experimentally. From detailed simulations, it was concluded that a cladding layer with an index close to SiO_2 (index~1.46) on top of the microring hardly affects the bending losses of the MR-mode.

This is confirmed by measurements on a microring resonator like in Figure 3 (left), but now with a PMMA cladding (index~1.476) on top. The spectral response, see Figure 4 (right), is similar to the microring resonator with air cladding, but the experimental finesse is somewhat lower: *F*~61. We attribute this mainly to the absorption of the MR mode in the PMMA layer. The overlap of the MR-mode with the PMMA cladding is however only 23%.
As an alternative we realized vertically coupled cylindrical microdisk resonators allowing for an independent design of the microdisk and the underlying (phase-matched) SWs. The coupling between the MR- and the SW-mode is controlled by the thickness of the separation layer (h_{sep}), which can be controlled accurately.

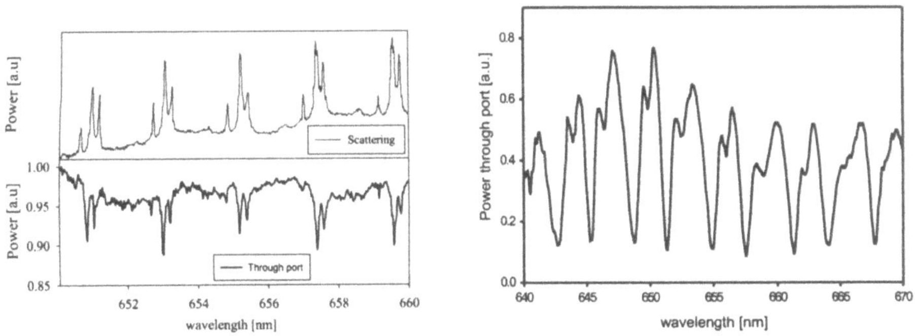

Figure 5. Measured spectra of a vertically coupled microdisk resonator with (left) Si_3N_4 core (see also Figure 3) and polyurethane core layer (right) respectively.

First consider the vertically coupled microdisk resonator with Si_3N_4 core, see also Figure 3. The measured dips and peaks in the through port and the power scattered by the microdisk respectively are consistent with one another (see also Figure 5). The microdisk supports three different whispering gallery modes (WGMs) whose 'experimental' finesse are limited by the linewidth of our dye laser: $F>31$. Using the calculated bending losses of the three WGMs in conjunction with the experiments we obtained a lower limit for the unloaded (i.e., κ=0) finesse and maximum G of the fundamental TE polarized WGM: $F_{un}>105$ and $G_{max}>5.4$.
We also realized a vertically coupled microdisk resonator with polyurethane (PU) core (index~1.55). Even though we did not optimise the device we can clearly observe two dips (see Figure 5) corresponding to the fundamental and first order WGMs in the microdisk. For the 'experimental' finesse of the fundamental MR we found *F~5*. The feasibility of vertically coupled microdisk resonators with a low index polymer core (which has an almost perfect overlap with the fundamental MR mode) is however clearly shown. By an optimised design we however expect similar a similar intra-cavity power enhancement as for the MRs with polymer core.
An overview of the experimental result is given in Table 1. From this table it can be concluded that both the lateral and vertical coupling schemes allow for high finesses and an enhancement of the intra-cavity power of at least one order of magnitude.

TABLE 1. Overview of the experimentally determined properties of the fundamental MR-mode. *) Finesse of an unloaded MR.

Core material	Upper cladding	Coupling scheme	Finesse	G_{max}
Si_3N_4	Air	Lateral	136	12.5
Si_3N_4	PMMA	Lateral	61	5.5
Si_3N_4	Air	Vertical	>105*)	>5.4
PU	Air	Vertical	~5	~0.6

4. Conclusions

The requirements for efficient all-optical switching have been investigated both theoretically and experimentally. High finesse laterally and vertically waveguide-coupled MRs with a maximum intra-cavity power enhancement of one order of magnitude have been demonstrated. Vertically coupled MRs are the most promising as they allow for an independent design of the MR and the SWs. We also demonstrated microdisk resonators with a low index polyurethane core, which have an almost perfect overlap with the MR-mode. By further optimising our MRs and using a polymer with Kerr-type non-linearity as core layer, we expect FWHM switching powers of only a few 100 mW.

5. References

[1] F.C. Blom et al., Experimental study of integrated optics microcavity resonators: towards all-optical switching devices. *Appl. Phys. Lett.*, vol. 71, pp. 747-749, 1997.

[2] A.F.J. Levi et al., Directional light coupling from microdisk lasers, *Appl. Phys. Lett.*, vol. 62, pp. 561-563, 1993.

[3] B.E. Little et al., Microring resonator channel dropping filters, *J. Lightwave Technol.*, vol. 15, pp. 998-1005, 1997.

[4] F.C. Blom et al., A single channel dropping filter based on a cylindrical micro-resonator, *Opt. Comm.*, vol. 167, pp. 77-82, 1999.

[5] B.E. Little et al., Microring resonators arrays for VLSI photonics, *IEEE Photon. Technol. Lett.*, vol. 12, pp. 323-325, 2000.

[6] D.J.W. Klunder et al., High-resolution photon-scanning tunneling microscope measurements of the whispering gallery modes in a cylindrical microresonator, *IEEE Photon. Technol. Lett.*, vol. 12, pp. 1531-1533, 2000.

[7] B.E. Little et al., Ultra-compact Si/SiO2 micro-ring resonator channel dropping filter, *IEEE Photon. Technol. Lett.*, vol. 10, pp. 549-551, 1998.

[8] D.J.W. Klunder et al., Design and characterization of waveguide-coupled cylindrical microring resonators in Si_3N_4, in: *Proc. LEOS 2000*, paper ThI2, 2000.

[9] Driessen et al., All-optical integrated optic devices: A hybrid approach, *IEE Proc.-Optoelectron.*, vol. 145, pp. 227-235, 1998.

[10] K. Wörhoff et al., Design, tolerance analysis, and fabrication of silicon oxynitride planar optical waveguides for communication devices, *J. Lightwave Technol.*, vol. 17, pp. 1401-1407, 1999.

PHOTONIC BAND GAP STRUCTURES IN PLANAR NONLINEAR WAVEGUIDES: APPLICATION TO SECOND HARMONIC GENERATION

M. BERTOLOTTI, C. SIBILIA
INFM, Dipartimento di Energetica, Università di Roma "La Sapienza"
Via A. Scarpa 16, 00161 Roma, Italy

Abstract

The properties of photonic band gap structures are discussed in view of their applications to second harmonic generation.

In integrated optics, waveguides offer the advantages of lateral confinement of the field and mode dispersion for phase-matching adjustment using different polarisations for the fundamental and second harmonic fields [1]. However, inter-modal dispersion generally leads to poor mode overlap in the transverse direction, thus making this phase mismatch compensation technique convenient only for materials with large nonlinear coefficient [2].

The use of a medium with periodic linear properties in the form of photonic band gap (PBG) structures can present some advantages. The linear properties of photonic band-gaps (PBG's) can be used to design efficient second-order interaction schemes. The main properties of interest for this kind of application in bulk are the possibility to use the dispersion of the structure to obtain perfect phase matching and the increased density of modes (DOM) that the structures exhibit at transmission resonance peaks when the spatial pulse length is much longer than the length of the structure. These properties have been discussed in a number of papers [3] and we will only briefly discuss them.

Fig. 1 shows the DOM as a function of frequency with the transmission spectrum superimposed, for a structure made alternating two layers of refractive index $n_1 = 1$ and $n_2 = 1.5$ and thicknesses $d_1 = \lambda_0/4$ respectively with $\lambda_0 = 1\,\mu m$. Simple inspection shows that the DOM has maxima at the frequencies corresponding to peaks of transmission. The DOM maximum is larger corresponding to the first transmission peak before a stop band. At the successive transmission peaks away from the stop band, DOM exhibits maxima which decrease with the position of the peak moving away from the stopband. The effective nonlinear coefficient can be shown to be proportional to the DOM [4] and is therefore higher at transmission peaks.

The second very interesting property of the structure is the possibility to add to the dispersion of the material with which the layers are made, the dispersion due to the geometry of the structure (form dispersion). Because in general the material dispersion increases with increasing frequency while the form dispersion decreases, it is possible in

A. Driessen (ed.), Nonlinear Optics for the Information Society, 63–67.

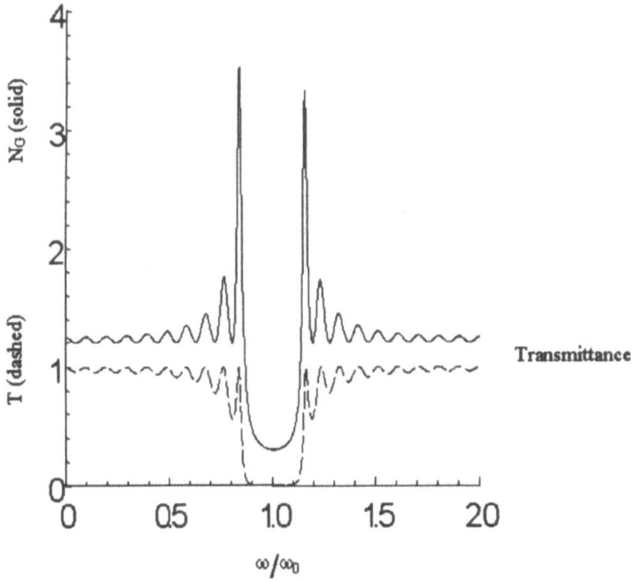

Figure 1: DOM(N_G) (upper curve) and transmittance (T) (lower curve) as a function of the normalized frequency.

a number of situation to compensate for material dispersion realizing perfect phase-matching. It has been shown [4] that the structure dispersion can be easily derived from the phase of transmission as can be calculated, for example, by using the transfer matrix formalism (6). An effective index can be derived for a finite structure (composed by a finite number of layers) so that the phase matching condition corresponds to equalize the effective indices for the fundamental and second harmonic wave in correspondence of peaks of transmission, so that

$$n_{eff}(\omega) = n_{eff}(2\omega) \tag{1}$$

Relation (1) can be expressed using the Bloch phase $\beta(\omega)$ corresponding to the infinite structure (7) according to the relation

$$2\beta(\omega) = \beta(2\omega) \tag{2}$$

It is therefore very easy to design structures which maximise phase-matching in correspondence of peaks of transmission where also the DOM has maxima.

A few examples can show how to use this strategy. Fig. 2 shows the situation that one must realize to have efficient second-order interaction. The transmission spectrum is shown which has the first maximum at the frequency corresponding to the fundamental frequency and the second maximum before the second stop-band at the frequency corresponding to the SH wave. Efficient SH generation can be predicted in this case. An experiment has been performed in which a GaAlAs/AlAs structure has been designed to have a reasonable $\chi^{(2)}$ coefficient growing the crystal in the (100) direction and

operating it with radiation incoming under an angle 10° with respect to the normal to the structure. It was not possible to compensate exactly the material dispersion, however an enhancement of efficiency was predicted and experimentally observed (8).
The great advantage of the use of PBG structures is that one may obtain very efficient interactions in very short structures.

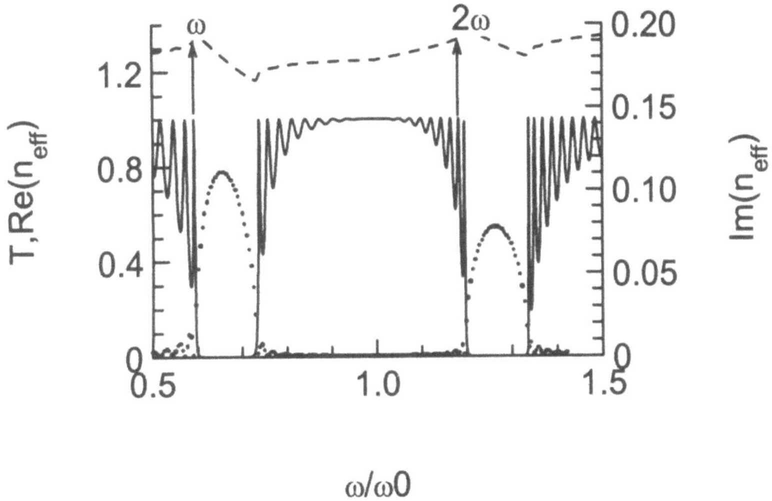

Figure 2: Transmission (T) vs normalized frequency (full curve) together with the real (dashed) and imaginary part (dotted) of the effective index.

Figs. 3 show the predicted efficiencies for the case of production of second harmonic from 1μm radiation to obtain green emission (fig. 3a) or from 0.8μm to have blue radiation (fig. 3b) as a function of input power for a PBG structure made alternating layers of AlN/SiO_2 or AlGaN/GaN respectively [8]. It is worth to observe that 15% or 5% efficiency is predicted in a single pass with a fundamental input power of about 4 GW/cm^2 in a structure as short as 4μm or 6μm, for green or blue production respectively.
Other second-order interactions have been studied as cascading nonlinear phase changes (10) or parametric generation (11) and other efficient interactions can be considered.

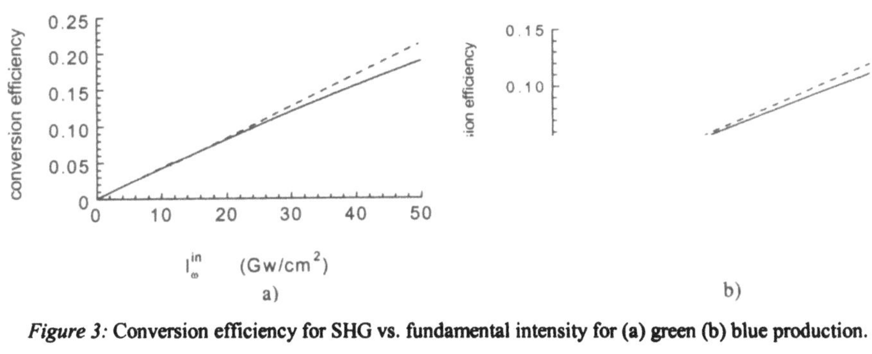

Figure 3: Conversion efficiency for SHG vs. fundamental intensity for (a) green (b) blue production.

In the case in which the pulse is so short to be less extended than the physical length of the device, it is no more possible to use the multiple reflections at the different interfaces to obtain space localization, but form-dispersion can still be used to increase efficiency in the structures (12).
When applying the PBG advantages to implement integrated devices in the form of waveguides the major distinction between a SHG process in a waveguide and in a PBG structure is that in the latter case it is possible to use pulses as short as a picosecond or less [3]. In the guided case, the narrow linewidth that the SHG process imposes on the phase-matching condition severely limits the frequency bandwidth of pump pulses that are used to feed the nonlinear device [11].
The simplest type of PBG is a one-dimensional, periodic or quasi-periodic multilayer stack [12], which has to be reproduced in a guiding substrate. The essential feature of the PBG structure is the index contrast between the different layers which form the stack. The greater is the index contrast, then the greater is the possibility of achieving both high DOM and phase-matching through a careful choice of the structure's parameters [4], i.e. number of periods, layers' thickness, angle of incidence, etc.
Waveguide electromagnetic propagation is characterised by the introduction of a guided index and a transverse field profile [13]. The mode propagates as in a bulk medium whose refractive index equals the guided wave refractive index. The transverse field profile is given by the solution of an eigenvalue problem: if the geometry of the structure has no imperfections, such as rough interfaces, then the solution remains the same along the propagation path.
The presence of the PBG allows to modify the effective index. This can be done either making the waveguide with a PBG structure or changing the depth of the grating or the index of the cladding index by making a grating in the form of a PBG. The change of the depth of the guide can be obtained by replacing the air on the film with a different cladding, i.e. a cladding with a refractive index as high as possible. Changing the guide's depth is equivalent to replace air with a cladding made of the same material. This is probably the easiest way of getting an artificial stack of different materials in a waveguide structure. The stack is artificial because the electromagnetic field does not really cross alternating layers of different materials. Rather, it propagates in the same medium which is slightly perturbed by the surface grating.
A high index contrast between alternating layers in the bulk case provides a high DOM when the fundamental field is tuned at the band edge resonance, and phase-matching in terms of the effective index of refraction [4]. Phase-matching occurs when the second harmonic field is tuned at the second band edge resonance away of the second order stop band.: again the greater is the index contrast, the easier it is to arrive at phase-matching conditions by using a small number of periods and even normal incidence.
The index contrast achievable in commonly used waveguides for nonlinear devices is limited by the contrast between the refractive index of the film and the refractive index of the substrate. To be more specific, we may consider a waveguide made by using a $LiNbO_3$ substrate, because of the well-established growth and fabrication techniques and the high nonlinear coefficient typical of this material. Depending on the process used to grow the guiding layer on the substrate, one can get a range of index steps between the guide and the substrate. Metal in-diffusion techniques generally lead to Gaussian index profiles, with maximum index modulation depth of order 10^{-2}. In

contrast, ion exchange techniques lead to step index profiles as high as 0.13. These changes are however still one order of magnitude smaller than the typical normal dispersion exhibited between fundamental and second harmonic field in $LiNbO_3$. The use of the equivalent PBG structure thus allows the introduction of a high DOM, but the phase-matching conditions must be obtained by different means, i.e., with the introduction of a linear or a nonlinear grating as in quasi-phase-matched devices. Another difficulty is encountered. The small index contrast achievable in the waveguide can produce a high DOM only at the fundamental field. However, if the phase matching condition is fulfilled anyhow by other means, the qualitative result for SHG conversion efficiency is the same as that obtained in PBG structures with deep gratings. The results of a study made through a full vectorial coupled mode theory which permits both linear and nonlinear mode coupling to be taken into account show that the SHG process may be enhanced in a PBG waveguide. However, the theory shows that the band edge tuning conditions and the phase-matching requirements cannot be satisfied at the same time, but the introduction of a periodically poled substrate may allow to achieve phase-matching over the entire length of the nonlinear device.

References

[1] R.A. Andrews, "Crystal symmetry effects on nonlinear optical processes in optical waveguides", *IEEE J. Quantum Electron.*, vol. 7, pp. 523-529 (1971)

[2] J.T. Boyd, "Theory of parametric oscillation phase matched in GaAs thin film waveguides", *IEEE J. Quantum Electron.*, vol 8, pp. 788-796 (1972)

[3] J.M. Bendickson, J.P. Dowling, M. Scalora, "Analytic expressions for the electromagnetic mode density in finite, one-dimensional, photonic band-gap structures", *Phys. Rev. E*, vol. 53, pp. 4107-4121 (1996)
M. Scalora, M.J. Bloemer, A.S. Manka, J.P. Dowling, C.M. Bowden, R. Viswanathan, J.W. Haus, "Pulsed second-harmonic generation in nonlinear, one-dimensional periodic structures", *Phys. Rev. A*, vol. 56, pp. 3166-3174 (1997)
J.W. Haus, R. Viswanathan, M. Scalora, A.G. Kalocsai, J.D. Cole, J. Theimer, "Enhanced second-harmonic generation in media with a weak periodicity", *Phys. Rev. A*, vol. 57, pp. 2120-2128 (1998)

[4] M. Centini, C. Sibilia, M. Scalora, G.D'Aguanno, M. Bertolotti, M.J. Bloemer, C.M. Bowden, I. Nefedov, "Dispersive properties of finite, one-dimensional photonic band-gap structures: applications to nonlinear quadratic interactions", *Phys. Rev. E*, vol. 60, pp. 4891-4898 (1999)

[5] M. Born, E. Wolf, "Principles of Optics", Pergamon Press – Fifthhead (1975)

[6] F. Bloch, Z. Phys. 52 (1928) p. 555
A. Yariv and P. Yeh, "Optical waves in crystals", J. Wiley & Sons, N.Y. (1984) ch. 6
L. Brillouin, "Wave propagation in periodic structures", Dover 1953

[7] Y. Dumeige, P. Vidakovic, S. Sanvage, I. Sagues, J.A. Levenson, C. Sibilia, M. Centini, G. D'Aguanno, M. Scalora, "Enhancement of Second Harmpnic Generation in a 1D semiconductor PBG" - submitted on APL

[8] M. Bertolotti, M.J. Bloemer, C.M. Bowden, M. Centini, G. D'Aguanno, M. Scalora, C. Sibilia, "Blue and green light emission: New directions and perspectives of application of one-dimensional PBG" – SPIE vol. 3944 (2000) pp. 680-687

[9] G. D'Aguanno, M. Centini, C. Sibilia, M. Bertolotti, M. Scalora, M.J. Bloemer, C.M. Bowden, "Enhancement of $\chi^{(2)}$ cascading process in 1-D PBG" – Opt. Lett. (1999) vol. 24 p. 1663

[10] M. Centini, M. Scalora, G. D'Aguanno, M.J. Bloemer, C.M. Bowden, J.W. Haus, C. Sibilia, M. Bertolotti, "Efficient Nonlinear Infrared parametric generation in 1-D PBG", in press on Opt. Comm.

[11] s. for ex.
V. Berger "Photonic Crystals for Frequency Conversion Phase Matching in Nanoscale and Nonlinear Optics" – eds. M. Bertolotti, C.M. Bowden and C. Sibilia
A.M. Zheltikov "Controlling Light Pulses and Light Beams with photonic band-gap structures", ibidem

PHOTONIC BAND GAP LOGIC ELEMENTS FOR OPTICAL DATA PROCESSING

I. NEFEDOV[1], Y. MOROZOV[1], V. GUSYATNIKOV[2], A. ZHELTIKOV[3]

1) *Institute of Radio Engineering and Electronics, Russian Academy of Sciences, Zelyonaya str. 38, Saratov, 410019, Russia, NefedovIS@info.sgu.ru*
2) *Saratov State University, Astrakhanskaya Street 83, Saratov, Russia, victor@mail.saratov.ru*
3) *International Laser Center, Physics Faculty Moscow State University, Vorob'evy Gory, Moscow, 119899 Russia, zheltikov@nls1.ilc.msu.su*

Keywords

Photonic band gap structures, optical switching, optical logic gates

Abstract

The change in the refractive index of GaAs due to the light-induced generation of nonequilibrium charge carriers is shown to substantially change the transmission of a one-dimensional GaAs/GaAlAs photonic band-gap structure, allowing low-threshold photonic-crystal optical logic gates to be created. Elementary logic operations of the Boolean algebra performed with nonlinear PBG structures, including the biconditional implication, modulo-two addition, Peirce's arrow, and disjunction, are demonstrated. We show how different logic operations can be performed with the same GaAs/GaAlAs PBG structure at different power densities of logic pulses. Transmission characteristics of this structure can be tuned not only by light, but also by an electric field, which makes it possible to use the same structure as an electrooptical switch and to create hybrid optically and electrically controlled logic gates.

Introduction

The possibility of shifting the edge of the photonic band gap (PBG) in one-dimensional PBG structures due to the nonlinear (intensity-dependent) additive to the refractive index of optical materials forming a PBG structure offers much promise for the creation of tunable optical switches, bistable devices, and logic gates. In particular, Scalora et al.

A. Driessen (ed.), Nonlinear Optics for the Information Society, 69–73.

[1] demonstrated the possibility of using the dynamic shift of the PBG edge for optical switching. Light-by-light control in semiconductor periodic GaAs-GaAlAs structures was considered in our recent work [2]. If the wavelength of the *controlling* light corresponds to intrinsic absorption edge of GaAs, then the refractive index is changed due to the light-induced generation of nonequilibrium charge carriers in this material. Parameters of the structure are chosen in such a way that the wavelength of the *controlled* light corresponds to the photonic band gap edge. Light-induced charge carriers give rise to a negative contribution to the permittivity of GaAs, resulting in a shift of the photonic band edge. Such a structure can be used as an optical switch. Considering the continuous regime of illumination, we have found that the power of controlling radiation depends on the lifetime of non-equilibrium carriers. For 40-period structure with the lifetime $\tau=10^{-7}$ s the optimal switching power density is equal to 6kW/cm^2. The response time of optical elements, which is generally determined by inertia properties of optical nonlinearities employed in these elements, is a crucial parameter for all light-with-light-control devices, eventually characterizing the usefulness of PBG switches and logic gates as short-pulse-control components. It is very important in this context to understand the basic features of transient processes accompanying the propagation of light pulses in PBG structures.
We have studied also how the pulses of controlled light are formed by the controlling light pulses, whose duration is comparable with the lifetime of carriers. It was demonstrated that the minimal switching energy density does not depend on the lifetime and is equal to 0.0005 J/cm^2. Thus, the proposed mechanism of optical switching requires moderate energy densities and can be used in low-threshold all-optical devices. Analysis of changes in the transmission and reflection spectra of a one-dimensional PBG structure in continuous regime has demonstrated that the basic operations of Boolean algebra, including inversion, disjunction, and Peirce's arrow, can be performed with such nonlinear PBG structures [3]. In this paper, we show how the same basic logic operations can be implemented in pulse regime.

1. Optical and electrooptical switching of the transmission of a PBG structure.

The physical model of optical and electronic processes in GaAs/GaAlAs PBG structure under controlling light illumination as well as the method of numerical simulation, were described in [2]. It was found that even a small deviation of the wavelength of controlling light from optimal λ_0 causes essential change of the light controlling action. In fact, any small change in electronic properties of semiconductor layers caused by some external action around λ_0 gives rise to similar effects. In this paper we consider a bias electric field as such an external action. The structure under study is presented in Figure 2. The thicknesses of the layers are chosen in such a way that quarter-wave conditions for these layers are satisfied at the wavelength 1.45 µm. Then the long-wavelength edge of the photonic band gap corresponds to λ=1.5 µm. A bias dc electric field, applied to a semiconductor, reduces the effective width of the electronic band gap due to the Frantz-Keldish effect [4]. This reduction $\Delta\varepsilon_g$ can be approximately expressed via the bias electric field strength E_c as

$$\Delta\varepsilon_g = \left(\frac{3|e|E_c\hbar}{4\sqrt{2m}} \right)^{2/3} \quad (1.1)$$

where e and m are the charge and the effective mass of an electron, respectively. Such a reduction of the electronic band gap changes the absorption coefficient of the controlling light. Hence, an electric field may control the action of the controlling light. The transmission spectrum for a 40-period GaAs/GaAlAs structure near the long-wavelength band edge for different strengths of the electric field is presented in Figure 2. The lifetime of nonequilibrium charge carriers τ is taken to be 10^{-8} s and the energy flux density of the controlling light is P=40kW/cm^2. The structure with the sizes H=L=100μm was considered.

Figure 2 demonstrates that the spectral characteristics are identical for different wavelengths of the controlling light λ_1 =0.8727 μm and λ_2 =0.8730 μm if some special values of the external bias electric field are chosen. If E_c=0 (curve 1), the electric field does not influence the structure transmission, since λ_1 =0.8727 μm and λ_2=0.8730 μm are beyond the intrinsic absorption edge of GaAs at room temperature, and the absorption coefficient of such a radiation is very small. As the applied electric field grows, the transmission of the structure lowers (curves 2,3,4). Since $\lambda_2 > \lambda_1$ and λ_2 lies farther from the electronic band edge, one should apply larger external voltage to block radiation at λ_2. There exists an optimum for the controlling value of U_c. A further increase in U_c decreases the steepness of the photonic band edge (curve 5). The optimal value of U_c depends on the wavelength of controlling light.

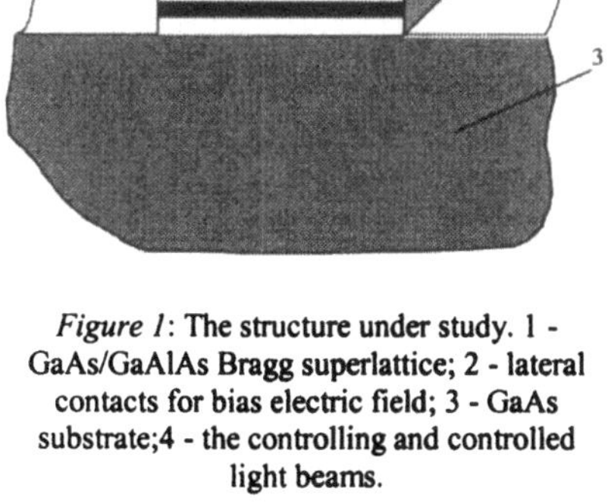

Figure 1: The structure under study. 1 - GaAs/GaAlAs Bragg superlattice; 2 - lateral contacts for bias electric field; 3 - GaAs substrate;4 - the controlling and controlled light beams.

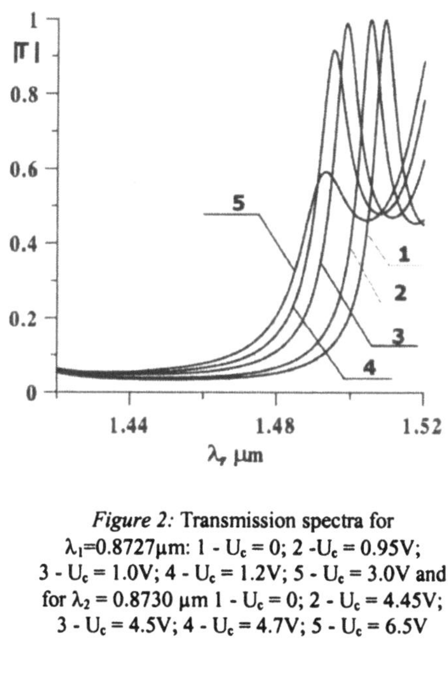

Figure 2: Transmission spectra for λ_1=0.8727μm: 1 - U_c = 0; 2 -U_c = 0.95V; 3 - U_c = 1.0V; 4 - U_c = 1.2V; 5 - U_c = 3.0V and for λ_2 = 0.8730 μm 1 - U_c = 0; 2 - U_c = 4.45V; 3 - U_c = 4.5V; 4 - U_c = 4.7V; 5 - U_c = 6.5V

2. Logic Gates

Logic gates based on PBG structures in the cw regime of illumination were considered in [3]. Below, we will show how the same logic operation can be implemented in the pulsed regime (see Figure 3 and Table 1). We assume that input pulses of controlling light having duration Δt form output pulses of the controlled light performing the logic functions. The truth table for logic functions NOR and OR can be constructed in the same way as in [3].

The sensitivity of transmission characteristics to the external electric field makes permits electrooptical switching. Next, let us consider a hybrid logic element, where controlling light is applied to Input 1 and an electric pulse is applied to Input 2 (see Figures 2, 4 and explanation in the Table 2). Importantly, the temperature changes of the electronic band gap width can be compensated by a small voltage electric field applied to the structure.

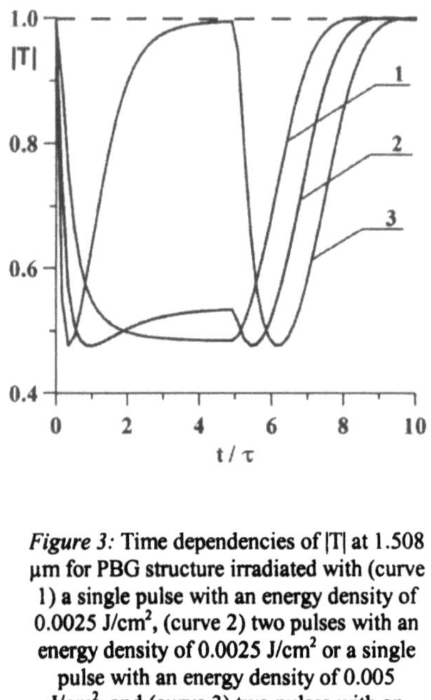

Figure 3: Time dependencies of |T| at 1.508 μm for PBG structure irradiated with (curve 1) a single pulse with an energy density of 0.0025 J/cm², (curve 2) two pulses with an energy density of 0.0025 J/cm² or a single pulse with an energy density of 0.005 J/cm², and (curve 3) two pulses with an energy density of 0.005 J/cm². Dash line corresponds to the absence of the controlling light.

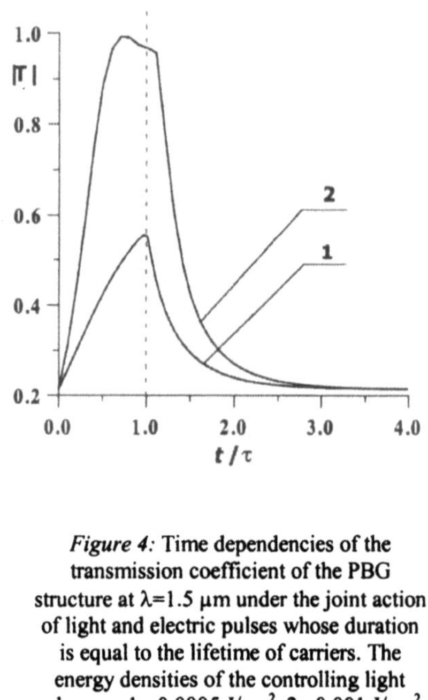

Figure 4: Time dependencies of the transmission coefficient of the PBG structure at λ=1.5 μm under the joint action of light and electric pulses whose duration is equal to the lifetime of carriers. The energy densities of the controlling light pulses are 1 - 0.0005 J/cm², 2 - 0.001 J/cm². The dash line corresponds to the moment of time when the controlling light pulse terminates.

TABLE 1: The truth table corresponding to the case when each of the 0.8725-μm logic pulses with a duration $\Delta t=5\tau$ switches the transmission of a PBG structure at 1.508 μm to the low level (as illustrated by curve 2 in Figure 3, corresponding to a single pulse with an energy density of 0.005 J/cm^2), while the joint action of two logic beams gives rise to a transmission maximum at 1.508 μm (as in the case of two light pulses with an energy density of 0.005 J/cm^2, curve 3 in Figure 3)

Input 1	*Input 2*	*Transmission*	*Reflection*
0	0	1	0
0	1	0	1
1	0	0	1
1	1	1	0
Function		Biconditional implication	Modulo-two addition

TABLE 2: The truth table for the electrooptical switching at λ = 1.5 μm corresponding to the curves 1,3, Figure 2, and curve 2, Figure 4. Input 1 corresponds to the action by light pulse at λ =0.8727 and energy density E=0.001 J/cm^2 on the structure. Input 2 corresponds to the action by electric pulse having amplitude 1~V to the lateral contacts of the structure. Logic zero corresponds to the cases when transmission or reflection coefficient are less than 0.5, and logic unit corresponds to these values close to 1.

Input 1	*Input 2*	*Transmission*	*Reflection*
0	0	0	1
0	1	0	1
1	0	0	1
1	1	1	0
Function		Conjunction	Dagger operation

3. Summary

Thus, we have shown the possibility of creating low-threshold all-optical and electrooptical logic gates, which can operate at room temperature. The inertia of the switching process is determined by the lifetime of carriers and was taken in our calculations from 10^{-7} to 10^{-9}s.

Acknowledgements

This work was supported by the Russian Foundation for Basic Research, Grants Nos. 00-02-17554, 00-02-81022, and 00-02-17567 and the CRDF projects RP2-2275 and REC 006.

References

[1] M. Scalora, J.P. Dowling, C.M. Bowden, M.J. Bloemer (1994). Optical Limiting and Switching of Ultrashort Pulses in Nonlinear Photonic Band Gap Materials. *Phys. Rev. Lett.*, *73*, 1368-1371.

[2] I.S. Nefedov, V.N. Gusyatnikov (2000). Optically controlled GaAs-GaAlAs photonic band gap structure. *Journ. of Optics A: Pure and Applied Optics, 2,* 344-347.

[3] I.S. Nefedov, V.N. Gusyatnikov, P.K. Kashkarov, A.M. Zheltikov (2000). Low-Threshold Photonic Band-Gap Optical Logic Gates. *Lasers Physics, 10,* 640-643.

[4] K. Zeeger (1973). *Semiconductor Physics* Springer, Berlin.

SILICON MICROPHOTONICS

P.G. KIK, M.J.A. DE DOOD, A. POLMAN
FOM-institute AMOLF
Amsterdam, The Netherlands

Abstract

Due to its mature processing technology, silicon is the ideal base material for microphotonic devices. First of all, it can serve as a breadboard for passive and active integrated optical components such as e.g. waveguides and optical amplifiers. Second the semiconducting properties of silicon itself can be exploited to fabricate novel opto-electronic devices. In this talk, three new concepts based in Si microphotonic technology will be adressed.

First, we will show the Si nanocrystals, embedded in a dielectric waveguide, can serve as an efficient sensitizer in an erbium-doped waveguide amplifier operating at 1.5 micron. The energy transfer between the nanocrystal excitonic system and the Er 4f shell will be discussed. Optical waveguides are made, based on the index contrast between the Si nanocrystals and the surrounding SiO_2, and optical gain measurements will be presented.

Next, a novel erbium doped Si infrared waveguide detector will be presented, that operates at 1.5 micron through an energy backtransfer process between the Er 4f shell and the Si electronic system, at an internal quantum efficiency as high as 70% at room temperature.

Finally, we will present a two-dimensional photonic bandgap structure at 1.5 micron made by lithography defining a cubic array of Si pillars, 5 microns tall and 270 nm in diameter, integrated with Si in-and output waveguides. A waveguide bend with a bending radius less than 2 micron will be shown, that may serve to achieve further miniaturizing in future Si microphotonic circuits.

Full article not available

A. Driessen (ed.), Nonlinear Optics for the Information Society, 75.

SEMICONDUCTOR OPTICAL AMPLIFIER DYNAMICS FOR WAVELENGTH CONVERSION APPLICATIONS

R. GUTIÉRREZ-CASTREJÓN, L. SCHARES, G. GUEKOS
Institute of Quantum Electronics, Swiss Federal Institute of Technology, Zürich,
ETH Hönggerberg, 8093 Zurich, Switzerland.
Tel. +41 1 6333773; Fax. +41 1 6331109; E-mail: gutierrez@iqe.phys.ethz.ch

Abstract

The data exchange efficiency of distributed optical information processing systems can be greatly spurred by the use of optical wavelength conversion (OWC). Based on the nonlinear optical effect of gain or absorption saturation, OWC can be carried out using the cross-phase or cross-gain modulation effects. Although OWC can in principle be performed in multi-level saturable-media, semiconductor optical amplifiers (SOAs) are preferred due to the broad spectrum, high gain, large nonlinearity, and fast effective carrier lifetime (τ) that they exhibit. The latter characteristic is particularly relevant since the conversion data sequence is limited by the speed of the SOA. In this contribution our measurements of τ in bulk InGaAsP/InP SOAs will be shown. They were performed at different levels of saturation on short and long amplifiers, where a τ as small as 60 ps for the long SOA was found. Additionally, a theoretical analysis of the experimental results in the light of a recently developed SOA model will be presented. Albeit its relative simplicity, our transient model takes into account a space and time-dependent carrier lifetime, the presence of the often disregarded amplified spontaneous emission and the nonlinear gain compression effect produced by carrier heating and spectral hole burning. The dynamical gain response of a SOA to a pulse sequence is also important for practical high-bit-rate applications. Therefore, corresponding curves at GHz data rates will be presented, and cross-phase modulation simulations that show the influence of τ as limiting factor on the performance of wavelength converters will be discussed.

Full article not available

A. Driessen (ed.), Nonlinear Optics for the Information Society, 77.

OPTICAL DATA STORAGE USING MICROHOLOGRAPHIC GRATINGS

S. ORLIC, H.J. EICHLER
Technical University Berlin, Institute of Optics

Keywords

optical memories, holography, Bragg gratings, multiplexing

Abstract

We present an overview of our research activities on optical data storage by following a so called microholographic approach that expands the conventional bit-oriented storage into the third dimension. Pits are replaced by reflection volume gratings created within a sub-micrometer range using highly focused laser beams. This article summarizes the main features of this storage method including system design and development, data coding and processing scheme as well as material testing.

1. Introduction

In conventional optical data storage technologies, the information is imprinted as quarterwave grooves in the reflective layer of the disk. The data is read out by projecting a focused laser spot on the rotating disk, then detecting the reflected light. As quarterwave surface steps lead to reflectivity changes by interference, the detected signal fluctuates with the presence or absence of pits along the track corresponding to the bit-wise stored information. The storage capacity is limited by the optical resolution of the pick up laser system. Data in currently available optical disks, especially in a DVD format, are already recorded close to the resolution limit. The implementation of blue-violet laser diodes will lead to an increase of the storage capacity by the factor 2.5 so that, for example, blue DVDs will have between 10 and 15 GB capacity depending on configuration. However, conventional optical storage is approaching physical limits so that new advanced concept are needed to overcome the storage density restrictions.
Holographic data storage is a well known volumetric approach that offers an attractive way to realize high capacity high data rates optical memories. Storing the information through the volume of a medium – not just on its surface, writing and reading an entire page at once, and storing multiple pages overlapping – are the main features of holographic storage. Such systems are being under extensive development since 1990. High storage densities and high data rates have been

A. Driessen (ed.), Nonlinear Optics for the Information Society, 79–87.

successfully demonstrated but present systems are still too complex and too expensive.
Therefore we propose an optical data storage system that combines the bit-oriented storage of conventional optical disks and volume storage from the holographic approach. The digital data is stored bit-wise as microscopic reflection gratings. The bit-oriented storage is expanded into the third dimension by recording the data holographically in localized volume elements. The storage medium is a thin photopolymer layer coated onto an optical disk substrate. The microholographic storage concept benefits from both technologies: The bit-oriented storage allows for using many solutions of the highly developed CD/DVD technology. On the other hand, holographic recording offers the way to overcome the limitations of this technologies related to its two-dimensional feature.

2. Microholographic Storage System

On a microholographic disk the pit-land structure of a CD is replaced by volume gratings which allow the bit-oriented storage to use the third dimension, i.e. the depth of the storage medium. For writing, microscopic reflection gratings are holographically induced to vary the reflectivity of the disk locally. A laser beam is focused into the photosensitive layer and reflected back with the mirror. The interference pattern of the incident and reflected beam induces a gratinglike modulation of the refractive index of the storage medium.

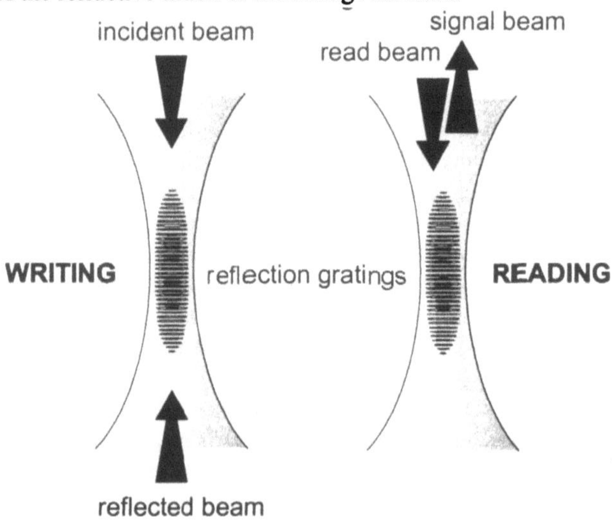

Figure 1. Microgratings are created in the beam waist of two counterpropagating focused laser beams. For detection, the read beam is diffracted at the induced gratings to reconstruct the original signal beam.

To retrieve the stored data, the original signal beam is reconstructed by reflection of the read beam at the induced gratings. Recording with sharply focused laser beams results in localized volume storage. The microgratings can be packed very densely and arranged in tracks similar to those on a CD. When the disk is rotating, microgratings of variable length are induced dynamically whereby grating fringes are extended in the motion direction.

The microholographic storage system is very similar to the conventional recordable or rewritable optical disk systems: A focused laser beam is used for writing and reading, the data is stored bit-wise in tracks on the rotating disk, similar systems for automatic focusing and track following are needed to control the position of the laser beam focus on a rotating disk. All these common aspects simplify the practical realization of the microholographic system as many components developed for CD/DVD technology can be directly used or adapted for this purpose. The main difference here is the reflecting unit underneath the disk that is needed for writing. During readout, this unit, in the simplest realization an aspherical mirror, can be tilted or removed so that only reflection from the gratings will be detected. Multiplex and demultiplex units which provide recording and readout with several beams of different wavelength simultaneously will be completely new.

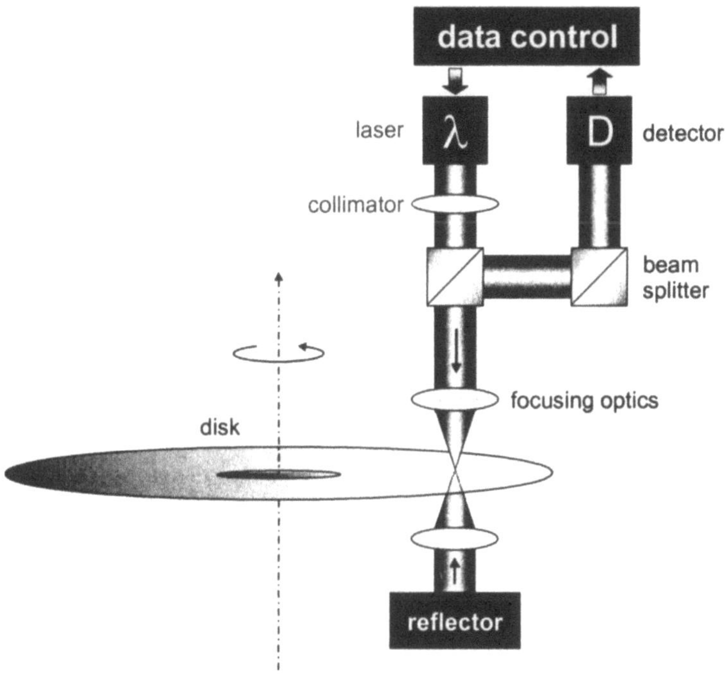

Figure 2. Schematic setup used for data recording. For writing, a laser beam is focused within a photosensitive layer and then reflected back from the mirror. The stored data is read out by detecting the local reflectivity of the disk; the reflecting unit is removed in this case.

3. Writing and Reading of Microgratings

Microgratings are written in the beam waist of two counterpropagating focused laser beams, one is incident and the other is reflected from the mirror. By the interference of these two beams, an intensity pattern is created that induces a corresponding spatial modulation of the refractive index of the storage medium. In general, it could be a modulation of the complex refractive index but photopolymer materials used for optical storage are predominantly pure phase media where only a modulation of the normal refractive index is considered. This periodic modulation represents a three-dimensional optical grating. The intensity pattern of two counterpropagating Gaussian beams calculated for DVD-like focusing conditions is depicted in figure 3. An effective modulation range, usually defined by the Gaussian envelope, is confined to the focal range of the laser beam. A periodic gratinglike modulation in longitudinal direction is rapidly decreasing with the distance from the write beam waist.

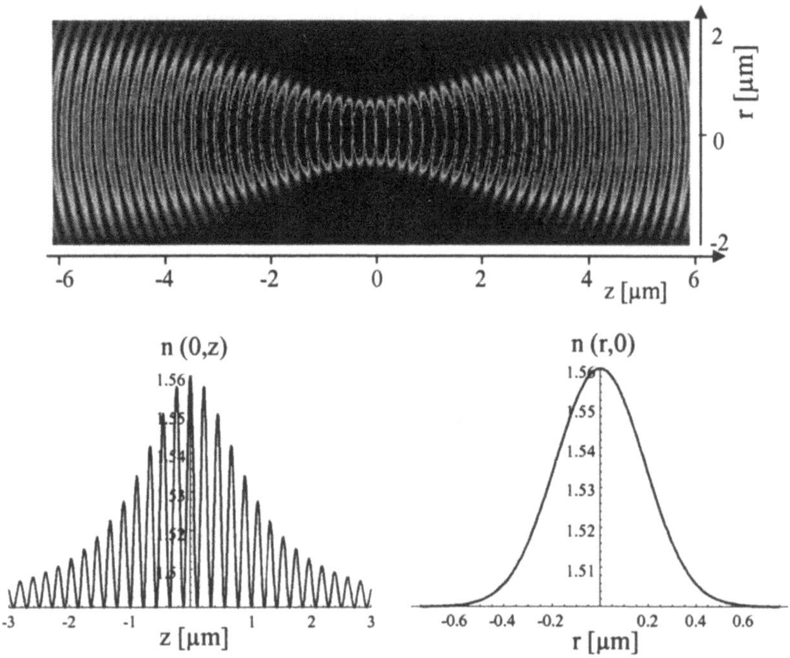

Figure 3. Calculated refractive index grating recorded by two counterpropagating Gaussian beams for DVD-like focusing conditions ($w_0 = 0.37$ µm, $\Delta n = 0.06$).

For modeling the reconstruction of the original information during readout, Bragg diffraction of a Gaussian beam from the volume grating shown before has been analyzed. For that, three-dimensional coupled wave equations, which describe the coupling of the incident read beam and reflected signal beam by an information bearing grating, have been used. The diffraction efficiency depends on the grating modulation, i.e. the maximum achievable refractive index change as well as on the thickness of the storage layer and on the wavelength of the laser light. Figure 4 shows the spectral response of a micrograting, i.e. the diffraction efficiency in dependence on the wavelength of the read beam. The diffraction efficiency of a micrograting is calculated as a function of the wavelength and is compared with this of an infinite holographic grating obtained using the plane-wave analysis. This comparison shows that microgratings also allow high diffraction efficiencies, even for thin storage layers (10 – 20 micron thickness). Being Bragg gratings, the microgratings are both wavelength and angle selective. The diffraction efficiency is rapidly decreasing with a deviation from the Bragg condition. Due to the Bragg selectivity, several microgratings can be stored overlapping in the same volume element and read out separately using laser beams of different wavelengths or directions. The selectivity of a grating depends on both the thickness of the material and the maximum refractive index modulation.

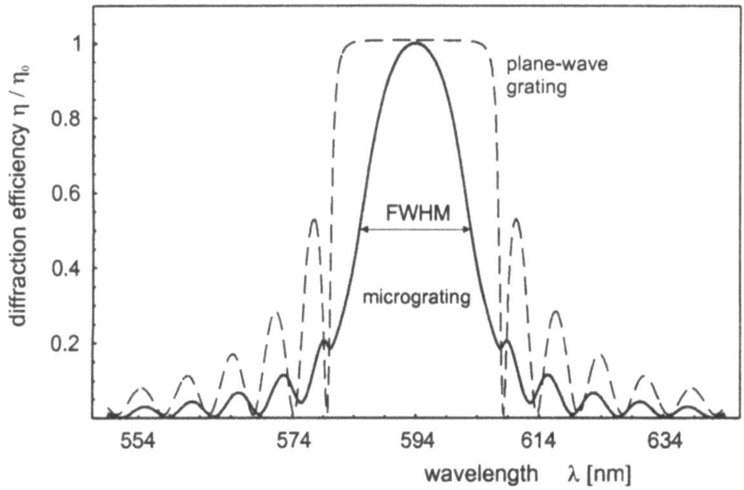

Figure 4. Diffraction efficiency of a grating as a function of the wavelength of the read beam. Wavelength selectivity of a grating depends on both the thickness of the storage layer and the maximum achievable refractive index modulation. For comparison, the efficiency of a plane-wave grating is calculated for the same parameters ($\Delta n = 0.06$, $d = 15$ μm).

4. Microholographic Multiplexing

The main advantages of microholographic storage concept compared with the conventional optical disk storage originate from both the usage of the whole volume of the storage medium and the application of holographic multiplexing methods. Various different multiplexing methods have been proposed for holographic data storage. Wavelength and angle multiplexing are resulting directly from the Bragg selectivity of holographic volume gratings but other multiplexing techniques such like phase, shift, peristrophic and fractal multiplexing can also be used. Multiplexing techniques offer an attractive way to increase both the storage density and data rates. Our concept concentrates on a combination of wavelength multiplexing and multilayer storage. Wavelength multiplexing is realized by simultaneous recording of several gratings in the same position with write beams of different wavelengths. In this case, a complex periodical grating structure is induced that contains all single-color gratings. Due to the wavelength selectivity, a single laser beam detects only the portions of corresponding gratings during readout. All bits contained in one multiplex grating can be recorded in parallel which increases the data rates, too.
In addition, localized volume storage also allows for spatial multiplexing which is realized as multilayer storage. Microgratings are recorded in different independent planes within the storage medium. A single data layer is addressed by the focused laser beam. Combining the wavelength multiplexing and multilayer storage, high storage densities can be achieved. Even moderate multiplexing factors would allow

to increase the storage capacity of a single disk with a DVD-specification to more than 100 GB.

5. Holographic recording in photopolymers

In general, a very important aspect in the development of new storage systems is the storage material itself. In the last years, a new class of photosensitive polymers have been introduced to satisfy the demand on adequate materials for holographic data storage. The recording mechanism is based on laser-induced polymerization. Diffusion of monomers separates the “bright” and “dark” regions. A light-induced grating-like modulation of the refractive index occurs during exposure and may be fixed after UV cure by polymerizing monomers in unexposed regions. In addition, holographic gratings recorded in DuPont´s photopolymers can be thermally processed to obtain higher efficiency.

In the field of photopolymers the progress in the last couple of years has been impressive. The main progress has been achieved by introducing new photopolymer films with large thickness of the order of 0.5-1 mm, such as Polaroid polymers. Further advances have been achieved, for example, Polaroid polymers are self-fixing so that an additional processing is not needed. An important aspect is the shrinkage of the material after exposure that has essentially been reduced to less than 0.1% in Polaroid polymers.

Testing and evaluating materials was and will remain an important part of our work. Several photopolymer materials have been characterized for microholographic storage including DuPont HRF, Polaroid polymers, azobenzene side-chain liquid crystal polymers as well as photorefractive polymers. Gratings recorded in azobenzene polymers can be erased by a circularly polarized laser beam so that rewritable storage becomes possible.

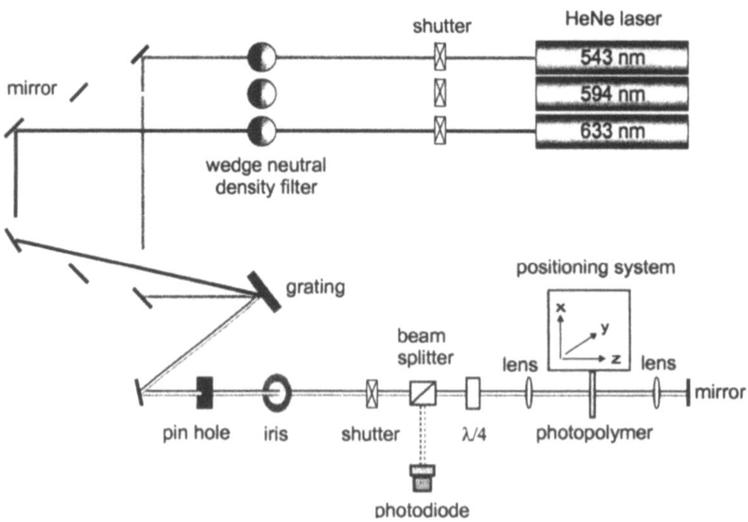

Figure 5. Experimental setup for wavelength multiplexing.

For material characterization but also for wavelength multiplexing, we use the experimental setup depicted in figure 5 that is equipped with three he-ne lasers emitting red, yellow and green light. The polymer sample is mounted on a three-axes translation stage equipped with piezo actuators. Both lenses are aligned to match the wavefronts of the counterpropagating beams in the focal region where the grating formation takes place.

In order to optimize exposure conditions, arrays of microgratings are recorded by varying both the laser beam intensity and the exposure time. The diffraction efficiency is measured spatially resolved by scanning the array with a focused probe beam and high positioning resolution of 10 nm only. In 15μm thick DuPont photopolymer films, high diffraction efficiency up to 80% has been obtained corresponding to the refractive index modulation of 0.05. Polaroid polymers with a thickness of up to 500 μm appear to be more suited as storage media although the photosensitivity is lower. The material is very homogenous and it is thick enough to allow the recording of multiple layers through the depth of the sample. DuPont polymers have the advantage to be full color holographic material that allows for wavelength multiplexing. The array of microgratings shown in figure 6 has been recorded with different wavelength combinations in DuPont photopolymer. The first three rows are single-color gratings, followed by three rows of two-color gratings and the last one is recorded with all three colors simultaneously. All colors are clearly resolved in the corresponding transmission spectra. The spectral width is less than 10 nm.

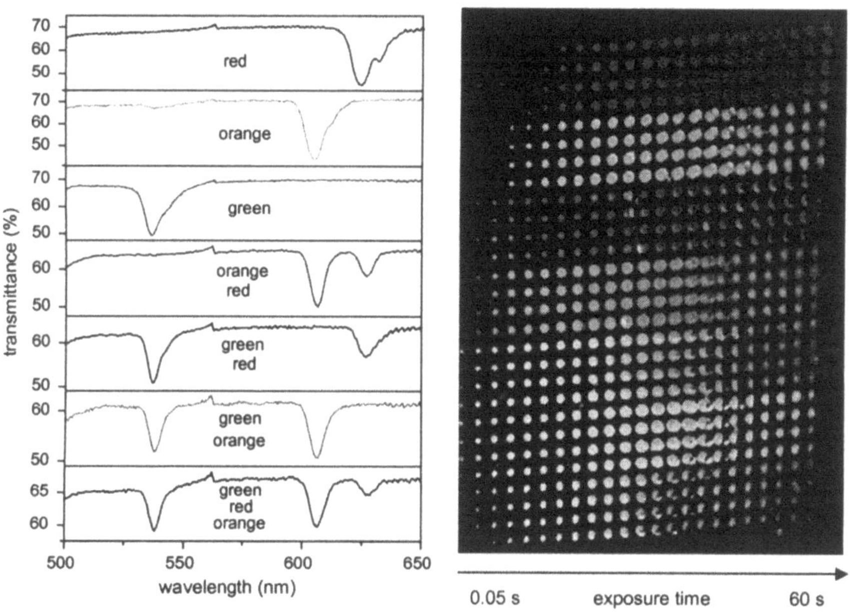

Figure 6. Array of microgratings recorded with different wavelength combinations in DuPont HFR 800 photopolymer film.

6. Coding and signal processing

Another crucial aspect for practical implementation of new data storage methods is coding and signal processing. Among other things, the success of the CD was enabled by powerful error-correction codes which reduce bit error rates appeared in the detection. As the microholographic concept is bit-oriented, the aims of coding are basically same as for CD/DVD. Both the physical dimension of the storing structure and the signal processing scheme will be very similar to CD/DVD. The data is, prior to recording, translated in two steps, error-correction code and recording or modulation code. The advantage of the microholographic method as compared to the page-oriented holographic memories, is that the bit-oriented storage allows for using the sophisticated coding algorithms evolved in CD/DVD technology. So the Cross-Interleaved Reed-Solomon code and the Eight-to-Fourteen Modulation seem to be optimized for microholographic storage systems, too.
To maximize the writing and reading speed the coding should be integrated parallel in the multiplexing scheme: The original data arranged in blocks is spread over multiplexing channels and coded parallel. Each of these channels corresponds to data channels in conventional optical systems so that error correction and modular coding can be realized using existing chips. By applying such coding scheme, even downward compatibility may be achieved.
On the other hand, specific requirements, which allows for applying established error-correction and modulation codes, are set on the physical structure of the data to be stored. The modulation code pursues the aim of structuring the data as well as maximizing the storage density and reliability. In a simple coding scheme each micrograting represents one bit. However, to realize an areal structure of the stored data similar to the pit-land structure on a CD/DVD, stripe-shaped gratings of variable length are required. In this case, binary 1s are represented as transitions along the track and the data is read out as the reflectivity change detected by the read beam. To make a profit of EFM, each channel stream has to be converted to a NRZI (non return to zero inverted) signal prior to physical recording. In this way the smallest microholographic unit represents 4 bits instead of 1 when using a simple coding scheme. EFM modulation will maximize the grating length by reducing frequent transitions. This presumes the recording of microgratings of variable length along the track whereby the grating length is determined by a coded bit stream.

To demonstrate this experimentally, the photopolymer sample in the setup depicted in figure 5 is moved with a constant linear velocity perpendicular to the write laser beam resulting in a dynamic formation of grating fringes in the motion direction. The grating length depends on the exposure time. An array of microgratings which have been induced dynamically as stripe-shaped volume gratings of variable length is shown in figure 7. Independent on the grating length, the modulation range is clearly confined in both lateral and motion direction.

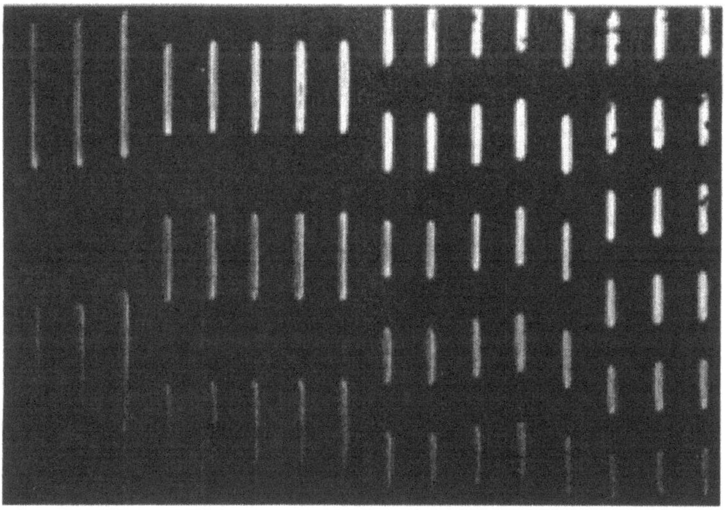

Figure 7. Dynamic writing and reading: Array of stripe-shaped microgratings of variable length recorded at 594 nm; the write beam radius was 10 µm.

7. Conclusions

To summarize the main features of the microholographic storage concept we can point out that the pit-land structure as applied in conventional optical storage technologies is replaced by microscopic reflection gratings. So the bit-oriented optical storage is expanded into the third dimension by recording the bit-formatted digital data as localized volume gratings. The Bragg selectivity of microgratings allows for applying holographic multiplexing techniques. The storage capacity of present optical disks may be surpassed by combining wavelength multiplexing and multilayer storage. The concept allows to take over many solutions of the well established CD/DVD technology. Our further research will follow three main directions. Decisive for the system development will be the implementation of blue laser diodes. Physical constrains have to be explored to find how small may the microgratings become to still be detected. Wavelength multiplexing has to be realized within the blue range which implies the investigation and optimization of the wavelength selectivity of microgratings by adjusting the exposure parameter. The storage material remains a crucial aspect – we are working with Polaroid on optimizing and further developing ULSH polymers but other materials such like azobenzene or photorefractive polymers may also become attractive as reversible storage media. And finally, coding and signal processing has to be implemented in a first working system. We expect that ongoing efforts will lead to a successful technological realization of this approach. Microholographic storage disks might be successful as removable storage media, which can satisfy high-capacity demands in specific data storage and processing areas including data banks, archiving and security systems as well as image processing and new multimedia applications.

HIGH RATE VANDER LUGT AND JOINT FOURIER TRANSFORM CORRELATORS

R. RYF[1], D. HAERTLE[1], G. MONTEMEZZANI[1], P. GÜNTER[1], A.A. GRABAR[2]

1) *Nonlinear Optics Laboratory, Swiss Federal Institute of Technology, ETH Hönggerberg, CH-8093 Zürich (Switzerland)*
2) *Institute of Solid State Physics and Chemistry Uzhgorod State University, 294000 Uzhgorod (Ukraine)*

Abstract

Compared to electronic systems, optical correlators are potentially faster because processing is performed in a single parallel step. Here we demonstrate correlation rates of 10.000 correlations/s using two different approaches.
The first is based on the Vander Lugt type correlator in which 100 holographic matched filters implemented in photorefractive crystals. Reading out the stored holographic memory with the new input image and detecting the most intense reconstructed reference wave delivers the correlation function. We compare the correlation fidelity for the cases where the images are in the form of intensity only and phase only spatial modulation, respectively. It is verified that the best quality correlation is achieved by coding the input scenes as binary phase images, in agreement with our expectations.
The second approach is based on a Joint Fourier Transform (JFT) correlator scheme. The core of the system is a real time holographic material ($Sn_2P_2S_6$) performing the correlation. This crystal exhibits interband photorefractive effects at 532 nm, which are advantageous for this task because of fast response (τ = 4 μs for a pulse fluence of 110μJ/cm^2) and thin holographic gratings, thus preserving the correlator shift invariance. The JFT-correlator is operated under pulsed illumination of a frequency double diode pumped Q-switched Nd:YAG laser at a repetition rate of 10 kHz. The library images are generated at this ferroelectric liquid crystals spatial light modulator (256x256 pixels). Successful correlation among 10 stored images is demonstrated at this rate.

Full article not available

A. Driessen (ed.), Nonlinear Optics for the Information Society, 89.

REWRITABLE OPTICAL STORAGE BY POLARIZATION HOLOGRAPHY

F. UJHELYI[1], E. LŐRINCZ[1], G. SZARVAS[1,2], J. FODOR[2], A. SÜTŐ[2], P.I. RICHTER[1], S. HVILSTED[4], P.S. RAMANUJAM[3]

1) *Budapest University of Technology and Economics, H-1111 Budafoki út 8., Hungary*
2) *Optilink Hungary Ltd., Graphisoft Park, 1031 Budapest, Hungary,*
3) *Risø National Laboratory, DK-4000 Roskilde, Denmark*
4) *Technical University of Denmark, DK-2800 KGS. Lyngby, Denmark*

Keywords

Holographic memory, polarisation holography, azobenzene polyester

Abstract

We present a novel solution for high-density rewritable optical storage of data in thin azobenzene containing polyester film. Experimental verification of reflected transmission polarisation holography for storing digital information is presented. Improved holographic system capable of achieving data density of 1 bit/μm^2 is under development. Hologram size of 0.06 mm^2 is achieved with a custom designed Fourier-objective.

1. Introduction

The rewritable holographic memory system of Optilink provides a novel solution for high-density optical storage. In contrast to most holographic storage systems that use highly multiplexed transmissive volume holograms in bulk materials [1], we use thin polyester layer on reflective substrate realising an optical arrangement first time in the world that allows writing and reading from the same side of the data carrier [6]. High data density is achieved in a parallel arrangement of the object and reference beams allowing also the use of different writing and reading wavelengths without optical distortion and realising in such a way a non-volatile storage. We use polarisation holography [7], e.g. recording is accomplished by combining object and reference beams polarised circularly in the opposite directions in the recording medium which is a photoaddressable polymer [5] allowing erasing and rewriting several times. Reflective data carrier has potential advantages, e.g. the substrate bulk optical quality is not important, handling is easier, double sided application is also feasible.

A. Driessen (ed.), Nonlinear Optics for the Information Society, 91–96.

2. Polarisation holography with reflective data carrier

We verified experimentally the possibility of storing digital information by polarisation holography when using 2 μm thick azobenzene containing polyester [2] on reflective glass substrate and parallel arrangement for the Fourier transform object beam and for the reference beam. In the laboratory set-up commercial SLM (Kopin 320x240 with 15 μm pixels) and polarisation elements (BS, PBS and waveplates) were used with a custom designed three-element Fourier-objective proving the feasibility of reflective transmission polarisation holography. The 9.5 mm focal length provides a theoretical data density of 0,16 bit/μm^2 and a hologram size of 0,67 mm square. Fig. 1 shows direct SLM image prepared in the laboratory system. The central area without binary data is maintained for the reference beam realising in such a way the parallel arrangement of the object and reference beams.

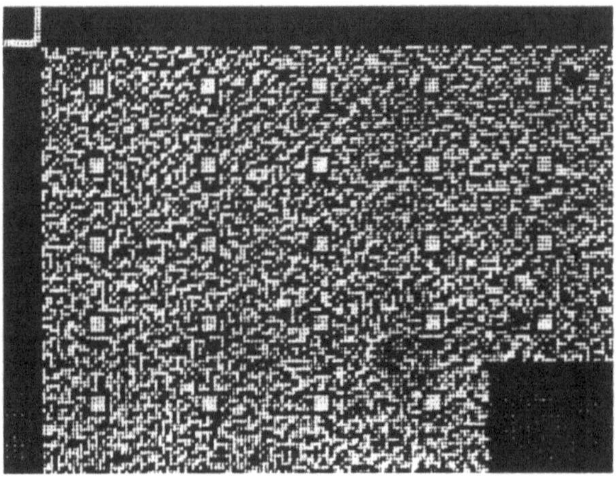

Figure 1 Readout of a quarter direct SLM page

Fig. 2 shows the retrieved image. We realised by defocusing a data density of ~0.1 bit/μm^2 and image reconstruction with a raw BER (Bit Error Rate without error correction) below 10^{-3} from image area beyond the white dashed line in Fig. 2. The central white spot arises from direct reflected part of the reference beam due to the commercial beam splitter and CCD used in the experiments.

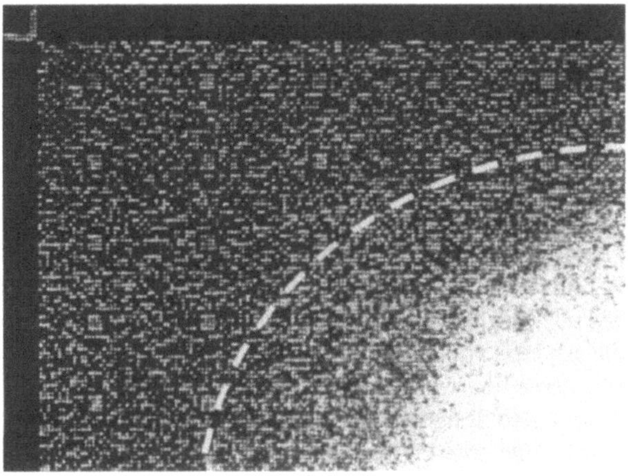

Figure 2 Recovered data page with the experimental system, data evaluation beyond white dashed line

3. Improved holographic system

Fig. 3 shows optical arrangement of the improved holographic system.

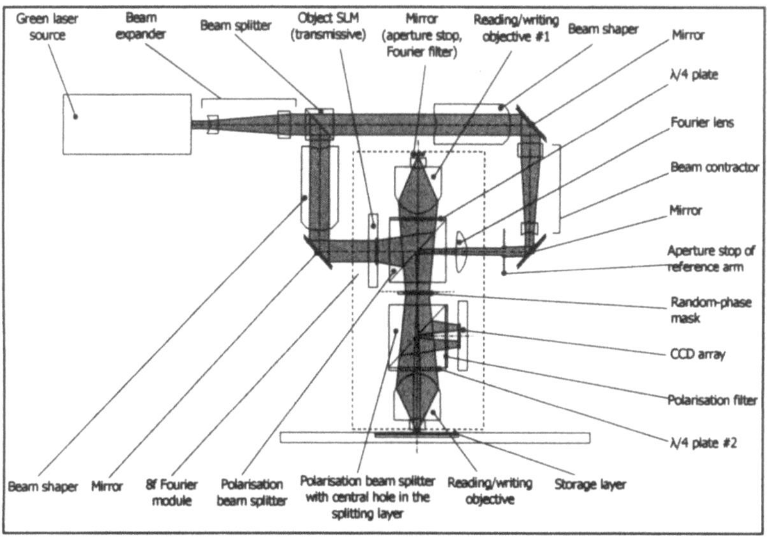

Figure 3 Optical arrangement of the improved system

A small frequency doubled Nd:YAG laser at 532 nm is the light source. The Fourier transformed data page is recorded in an 8f system containing two identical Fourier objectives and two polarisation beam splitters. Mirror with aperture stop in the 8f

module allows suitable Fourier filtering, while the random phase mask decreases undesirable intensity peaks in the Fourier plane. The plane reference wave arrives at normal incidence to the storage layer through the central hole of the second polarisation beam splitter resulting parallel object and reference illumination. During readout at the same wavelength, the major part of the reflected reference beam goes through the hole of the prism. This arrangement allows making a compact device.
We aimed a data density of 1 bit/μm^2 without multiplication that is higher then ever-realised [2]. It can be achieved by custom designed special optical elements. Two different Fourier objectives were planned and fabricated. A two element biaspherical-spherical lens had aberrations caused by out-of-tolerances surface slope [4] and could not be used in the system.
Fig. 4 shows the scheme of a working solution for the Fourier-objective, e.g. a three-element lens that was also designed and realised. A commercial aspherical lens was taken as a basis and reintegrated with two auxiliary spherical lenses. Parameters of the reading/writing objective are: object radius 2.98 mm, Fourier transform radius 0.154 mm, object NA 0.032, Fourier space NA 0.69.

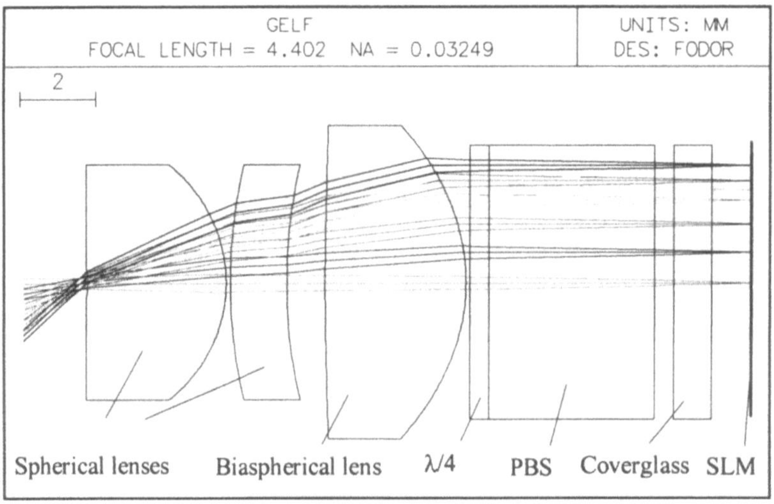

Figure 4 Scheme of working solution for the Fourier-objective: commercial aspherical lens reintegrated with two auxiliary spherical lenses

Fig. 5 shows result of an operational test made with the improved system.

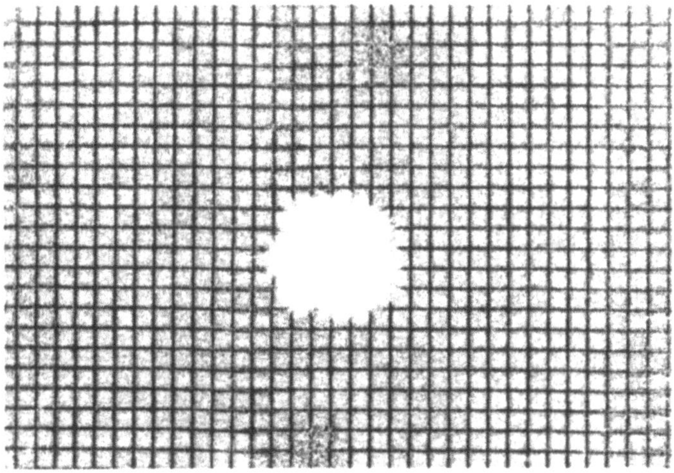

Figure 5 Direct SLM image of one-pixel grid for testing aberration in the optical system

The improved system stores information in small holograms on the reflective sample. The radius of the holograms is 0.14 mm. Fig.6 shows a recovered two-dimensional data page together with the error map after image reconstruction.

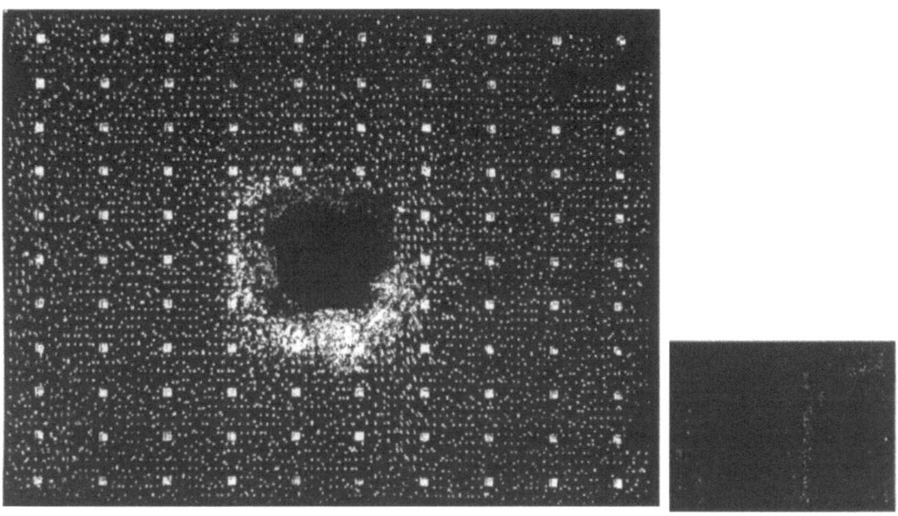

Figure 6 Reconstructed hologram in the improved system and error map after image reconstruction

Main sources of the errors are inhomogenity of the illuminating beams and polarisation layers, undesired scattering mainly in the centre area, inter-pixel crosstalk and the lack of pixel matching. Primary block organisation of data advantageously eliminates most of problems listed above. Actually, blocks of 30 x 20 pixels are used. Within each block a combination of pixels (a synchronisation code) allows localisation of the block and facilitates calculation of necessary co-ordinate transformation. Different methods of coding and signal processing have been presented by several authors including modulation and error-correction coding, thresholding, application of different decision schemes, pre- and post-processing of signals [1][8]. Necessary adaptation and improvement of known techniques are in progress. At present pages of 1.8 kbytes are written resulting in a data density of ~0.25 bit/μm^2.

4. Summary of the results

A polarisation holographic system is presented for high-density optical storage in thin photoaddressable polymer. We realised an experimental system first time in the world that is capable of storing the Fourier transform of digital information in reflective sample in a parallel arrangement. We also presented an improved holographic system containing out-of-the-common optical elements. Our three-element Fourier-objective provides a hologram size of 0.06 mm^2 that is smaller than ever realised and a data density of 1bit/μm^2 is feasible in our system. Unexpected amount of pre- and post-processing of signals restricts our system to be used presently at a reduced data density of ~0.25bit/μm^2.

References

[1] Burr, G. *et al.*, CLEO 1994.
[2] Burr, G. W. and Marcus, B, *Proc. of SPIE Vol. 3802*, p.18, 1999.
[3] Coufal, H. J. *et al.*, *Holographic Data Storage*, Springer, 2000.
[4] Erdei, G. *et al.*, will be published in *Proc. of SPIE Vol. 4093 B-67*, 2000
[5] Kerekes, A. *et al.* will be published in *Proc. of SPIE Vol. 4149*, 2000.
[6] Koppa, P. *et al.* WO 09957719A1, 1999.
[7] Todorov, T. *et al.*, *Appl. Opt. Vol. 23*, No. 24, pp. 4588-4591, 1984.
[8] Vadde, V. and Kumar, B.V.K.V. *Optics Letters Vol. 24*, No. 8., 1999.

NONLINEAR MAGNETO-OPTICAL PROBING OF ULTRATHIN MAGNETIC LAYER INTERFACES: CoNi/Pt AND Co/Cu

K. BAL[1], A. KIRILYUK[1], A. KEEN[1], K.J. VEENSTRA[1], TH. RASING[1], H.A.M. VAN DEN BERG[2], Y. LOU[3] K. SAMWER[3], M.A.M. HAAST[4], J. C. LODDER[4]

1) *Research Institute for Materials, University of Nijmegen, Toernooiveld 1, 6525 ED Nijmegen, The Netherlands tel:+31 24 3653094, fax: +31 24 3652190, kursatb@sci.kun.nl*
2) *Siemens AG, ZTMF I, Paul Gossenstrasse 100, 0-91052 Erlangen, Germany.*
3) *I.Physikalisches Institut, Gottingen University, Bunsenstrasse 9, 37073 Gottingen, Germany*
4) *MESA Research Institute, University of Twente, 7500 AE Enschede, The Netherlands.*

Abstract

The recently developed technique of Magnetization induced Second Harmonic Generation (MSHG) has been shown to be an ideal tool for the investigation of ultrathin magnetic multilayers [1] due to its extreme interface sensitivity with very large magneto-optical effects. We have used MSHG in combination with XRD, AFM, MR, and VSM techniques on CoNi/Pt (magnetic data storage) and Co/Cu (sensors and MRAM) multilayers to study their dependence on preparation condition and thermal stability:

(a) The crystallographic contribution of the nonlinear magneto optical response from CoNi/Pt interfaces appears to scale linearly with increasing interface roughness as determined by small angle X-ray scattering and Atomic Force Microscopy. From the MSHG magnetic contribution it follows that the increased interface roughness causes the interface moment to turn out of plane while the bulk of the film has an in-plane magnetization.

(b) The long-term thermal stability of giant magneto resistance and tunnel magneto resistance sensors found in magnetic multilayers with a hard artificial antiferromagnetic (AAF) subsystem has been investigated at 300°C. The annealing is found to cause a stabilizing increase in the transport signal, the saturation field of the antiferromagnetic coupling and the Co layer coercivity respectively and a decrease in the Co magnetization. The MSHG shows that the Co/Cu crystallographic and magnetic interface characteristics also display a similar behavior upon heating. The results

A. Driessen (ed.), Nonlinear Optics for the Information Society, 97–98.

indicate that relaxation of the film structure is responsible for the sudden change, while demixing of Co/Cu and granularization of the Co layer are taking place.

[1] Th. Rasing, J. Magn. Magn. Mat. 175, 35-50 (1997)

Full article not available

HYBRID SOLITON FIBRE LASER CHARACTERISATION

J.-E. P.C. COMMUNAL[1], G. LARGOUNEZ[2], A. B. GRUDININ[3], M. IN HET PANHUIS[1], W. J. BLAU[1]

1) *Material Ireland Polymer Research Centre, Department of Physics, Trinity College Dublin, Dublin2 Ireland.*
2) *Univerisitée Claude Bernard, Lyon 1. Current address: 11 Cours Vitton,69006 Lyon, France.*
3) *Optoelectronics Research Centre, University of Southampton, Southampton SO17 1BJ, UK.*

Keywords

soliton, erbium, fibre optics, autocorrelation, FROG, NALM.

Abstract

We study the pulses produced by an hybrid fibre laser design consisting of a Saturable Bragg Reflector and a Nonlinear Amplifying Loop Mirror. The laser is pumped by a 1 Watt diode pumped YAG. Mode-locking operation depends on the polarisation controller position. Once mode-locked the laser is stable at the fundamental frequency of 4MHz. The pulses are measured by Second Harmonic Generation autocorrelation. The traces are found to be consistent with the theory of trains of identical soliton pulses. These trains comprise 5 solitons of width (320±40) fs with a period of (3.3±0.1) ps.

1. Introduction

Soliton waves have attracted a lot of attention in the area of telecommunications. Their mathematical properties are well known and optical soliton are easily created from high power ultra-short pulses propagating in optical fibre. Another option to create solitons consists in using an erbium doped fibre amplifier in a laser design. This low power approach has the advantage of using already existing telecommunication technology.

2. Theory

Optical soliton arises as the combination of two phenomenons, Group Velocity Dispersion (GVD) and Self-Phase Modulation (SPM). The GVD is a linear effect,

A. Driessen (ed.), Nonlinear Optics for the Information Society, 99–103.

consequence of the chromatic dispersion. The SPM is a third order nonlinear effect depending on the nonlinear refractive index n_2. Solitons are solutions of the Nonlinear Schrödinger Equation (NLSE):

$$i\frac{\partial A}{\partial z}=\frac{1}{2}\beta_2\frac{\partial^2 A}{\partial t^2}-\gamma|A|^2 A, \tag{1}$$

where A is the amplitude of the pulse envelope, β_2 is the GVD parameter, and γ is the nonlinear coefficient responsible for SPM. Although Soliton solutions have been proved to exist in all types of GVD regimes, anomalous[1] ($\beta_2<0$), zero[2] and recently normal[3] ($\beta_2>0$), the simplest propagation case requires the regime to be anomalous. The first order soliton solution will then have the following form.

$$A(z,t)=\frac{1}{\tau}\sqrt{\frac{-\beta_2}{\gamma}}\cdot Sech\left(\frac{t}{\tau}\right)\cdot e^{-i\frac{z\beta_2}{2\tau^2}}, \tag{2}$$

where Sech is the hyperbolic secant function. In fused silica, the dispersion is normal below 1300nm and anomalous above. Erbium doped fibre lases at 1550nm in the anomalous regime. It's possible to modify the value of the GVD near zero-dispersion for 1550nm with dispersion shifted fibre. In this case, the third component β_3 of the dispersion is not negligible and a different form of the NLSE including this term has to be used. Although it doesn't appear in equation 2, as a consequence of the balance between GVD and SPM, solitons are positively chirped, and possess "wings" trailing and preceding them. And as it propagates in presence of amplification, like in a laser cavity, its "wings" will in turn amplify into solitons. This phenomenon takes place as long as the pulse propagates into the amplifier, continuously generating new pulses, thus creating a train of identical solitons.
The Nonlinear Amplifying Loop Mirror (NALM [4][5][6][7][8][9][10]) is usually used in an all-fibre laser design where two NALMs are opposed in a figure of 8[11]. The NALM length and the X fibre coupler ratio quantify the pulses energy. As a result, only solitons with suitable energy will be allowed to circulate the cavity and be amplified into mode-locked trains. The Saturable Bragg Reflector (SBR[12][13][14]) is used in an in-line configuration with the amplifier. The change in refractive index provides the switching necessary to mode-locking. Our laser is an hybrid[15] since it mixes the two methods; one end of the NALM is butt-coupled to the SBR, as seen on figure 1.

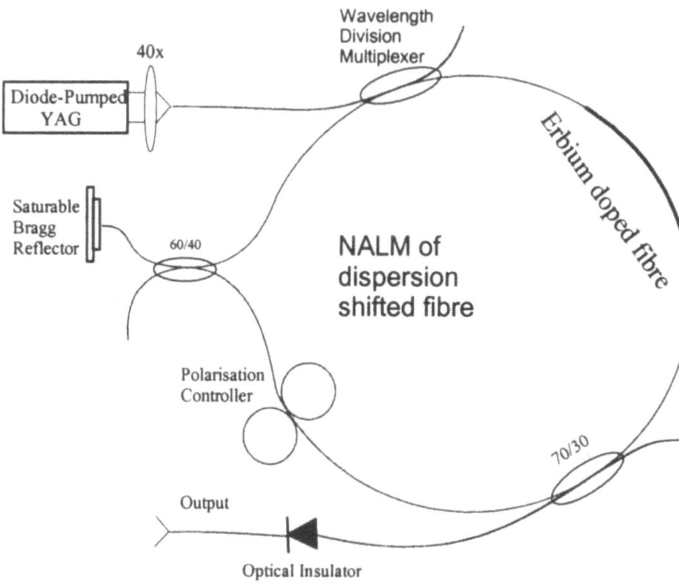

Figure 1. Hybrid fibre laser.

3. Experimental Set-up

We used CW diode-pumped YAG tunable up to 1W to pump the hybrid fibre laser. The laser output is coupled to the autocorrelator while the unused end of the NALM is coupled to an ultrafast InGaAs detector connected to a 300MHz analogue oscilloscope. This is necessary in order to tune the polarisation controller to achieve mode-locking, and to monitor the laser while the autocorrelation is being taken. The mode-locking frequency was thus measured to be 4MHz.

The autocorrelator is of classic design, (figure 2). The laser output first passes through a RG1000 filter. The first retro reflector is mounted on a high-resolution (100nm) computer controlled delay line. The recombined pulses are then focused with a lens of focal length 40cm on a Lithium Niobate Second Harmonic Generation (SHG) crystal. The SHG signal is then filtered by a monochromator and measured by a computer controlled CCD. This detection system was chosen to be easily upgraded from autocorrelation to Frequency Resolved Optical Gating (FROG).

4. Results

The best results were obtained by moving the delay line by steps of 10fs, and by cooling the to -30°C to reduce background noise. The number of solitons and the mode-locking frequency didn't vary with the pump power. This result contradicts the assumption that the number of pulses should increase with pump power since their energy is quantified. If it were the case, the number of pulses by train or the number of trains would increase. We observed neither, the variation in pump power could be insufficient to attain the

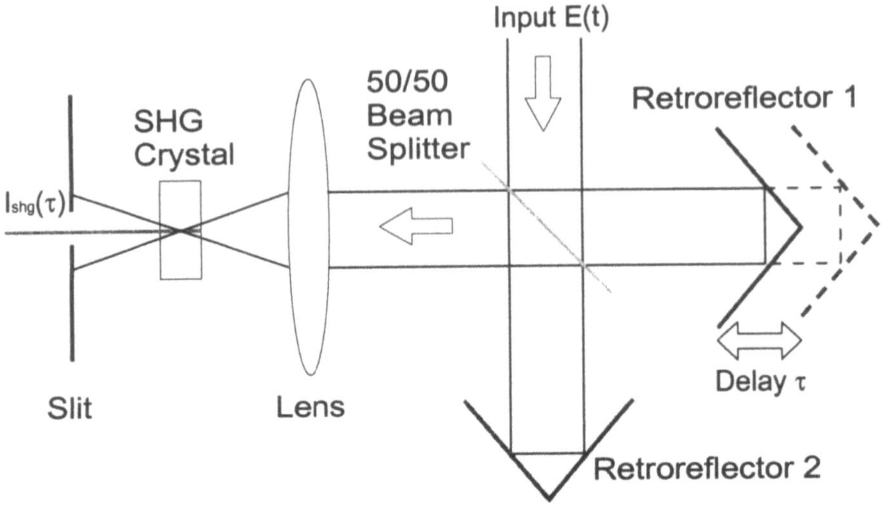

Figure 2. Autocorrelation set-up.

higher mode-locking harmonic. And since the number of pulses in a train is function of the length of the erbium doped fibre amplifier, the extra energy can go in the existing pulses who becomes higher order solitons. In this case, a soliton train would be made of pulses of different order, and this would be measured on the autocorrelation by different intensities and widths. This doesn't appear to be the case either. The traces measured (figure 3), are in excellent agreement with calculations based on a model of 5 identical solitons of width (320±40) fs at a period of (3.3±0.1) ps. The polarisation state was also measured to be elliptical, which would be consistent with a polarisation-locked solitons as previously observed with linear SBR laser design[13][16]. With an average power of 4mW, the peak power in the pulses is of the order of 500W, also in good agreement with the values given in the literature [1].

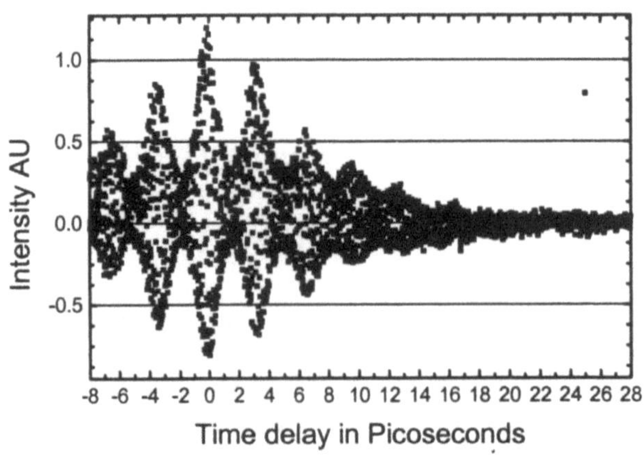

Figure 3. Hybrid fibre laser autocorrelation.

5. Conclusions

The results are in good agreement with the other existing systems. This laser design should allow us to go further modifying the cavity length by adding fibre near the SBR and thus changing the mode-locking frequency. Propagation in different length of fibre after the laser should also give us information on the soliton trains periodic evolutions. The implementation of FROG will allow us to measure the pulses chirp and complete the characterisation of this laser design.

Acknowledgements

We would like to thank the Irish Higher Education Authority (HEA) and the EU TMR Network Delos.

References

[1] G.P. Agrawal, *Nonlinear Fiber Optics.*
[2] G.P. Agrawal, and M.J. Potsek, *Phys. Rev. A 33*, pp1765-1776 (1986)
[3] N. Efremidis, K. Hizanidis, B.A. Malomed, H.E. Nistazakis, and D.J. Frantzeskakis, *J. Opt. Soc. Am. B 17*, pp952-958 (2000)
[4] N.J. Doran and David Wood, *J. Opt. Soc. Am. B 4*, pp1843-1846 (1987)
[5] N.J. Doran and David Wood, *Opt. Lett. 13*, pp56-58 (1988)
[6] M.N. Islam, E.R. Sunderman, R.H. Stolen, W.Pleibel, and J.R. Simpsonx, *Opt. Lett. 14*, pp811-813 (1988)
[7] K.J. Blow, N.J. Doran, and B.K. Nayar, *Opt. Lett. 14*, pp754-756 (1989)
[8] M.E. Fremann, F. Harberl, M. Hofer, and H. Hochreiter, *Opt. Lett. 15*, pp752-754 (1990)
[9] K. Smith, N.J. Doran, and P.G.J. Wigleyx, *Opt. Lett. 15*, pp1294-1296 (1990)
[10] J.D. Moores, K. Bergman, H.A. Haus, and E.P. Ippen, *Opt. Lett. 16*, pp138-140 (1991)
[11] J.W. Haus, G. Shaulov, E.A. Kuzin, and J. Sanchez-Mondragon, *Opt. Lett. 24*, pp376-378 (1999)
[12] J. Nathan Kutz, B.C. Collings, K. Bregman, S. Tsuda, S.T. Cundiff, W.H. Knox, P. Holmes and M. Weinstein, *J. Opt. Soc. Am. B 14*, pp2681-2690 (1997)
[13] S.T. Cundiff, B.C. Collings and W.H. Knox, *Optics Express Vol. 1 No. 1* (1997)
[14] N.N. Akhmediev, J.M. Soto-Crespo, S.T. Cundiff, B.C. Collings and W.H. Knox, *Opt. Lett. 23*, pp852-854 (1998)
[15] A.B. Grundini and S. Gray, *J. Opt. Soc. Am. B 14*, pp144-154 (1997)
[16] S.T. Cundiff, B.C. Collings, N.N. Akhmediev, J.M. Soto-Crespo, K. Bergman and W.H. Knox, *Phys. Rev. Lett. 82*, pp3988-3991 (1999)

PROGRESS IN HOLOGRAM WRITING IN Bi_2TeO_5 CRYSTALS

I FÖLDVÁRI[1] , C. DENZ[2], J. PETTER[2], F. VISINKA[2], Á. PÉTER[1]

1) *Research Institute for Solid State Physics and Optics, HAS, Konkoly-Thege u. 29-33, H-1121 Budapest, Hungary Phone: (36-1) – 3922627, FAX: 392-2223, e-mail: foldvari@power.szfki.kfki.hu*

2) *Institute of Applied Physics, Darmstadt University of Technology, Hochschulstr. 6, 64289 Darmstadt, Germany Phone: (49-6151) – 163-122, FAX: 164-123, e-mail: Cornelia.Denz@physik.tu-darmstadt.de*

Abstract

Choosing a proper volumetric memory material is a key question in holographic recording. Photorefractive bismuth tellurite crystals are a new candidate for this purpose. The crystal exhibits a long-living (more than 6 years) photorefractive signal component that develops in the four-wave mixing write process without any specific fixing.

In our contribution, we present the performance of Bi_2TeO_5 crystals that were grown by the Czochralski technique. Their photorefractive performance was tested by experiments investigating the four-wave mixing saturation diffraction efficiency. Moreover, two-dimensional images were written in Bi_2TeO_5 crystals in a typical holographic storage setup.

An improved crystal quality and the use of a stable Nd:YAG laser (532 nm) have led to an enhanced photorefractive response. The saturation diffraction efficiency of the written grating exceeded 40% in a given crystal orientation and light polarization (9 W/m^2 beams, 8 min saturation time, without external electric field). Holograms can be written with low intensity laser beams (data arm: 110 μW, reference arm: 680 μW) and with short exposure time (5 s). However, these conditions are not suitable for self-fixing of the image. Holograms written with longer exposures (1-8 min) are of good quality, with sharp contours and low background. Their modulation depth allowed observing the holograms by half an hour, thus approaching self-fixation processes.

Supports: COST Action P2, German - Hungarian joint R&D found, Hungarian Research Found OTKA (T-029756), Deutsche Forschungsgemeinschaft, DE 486/4-2.

Full article not available

A. Driessen (ed.), Nonlinear Optics for the Information Society, 105.

DISTRIBUTED RESPONSE ANALYSIS OF THE WATER DIMER

M. IN HET PANHUIS[1,2], R.W. MUNN[1], P.L.A. POPELIER[1], W.J. BLAU[2]
1) Department of Chemistry, UMIST, M60 1QD Manchester, U.K
2) Materials Ireland Polymer Research Centre, Department of Physics, Trinity College, Dublin 2, Ireland

Keywords

Distributed response, polarisability, *ab initio*, coupled perturbed Hartree-Fock, charge flow, water dimer.

Abstract

The understanding of the topological partitioning of electronic properties (TPEP) approach described in ref 6 was enhanced. Distributed polarisability calculations were employed to investigate the effect of interaction on electronic properties in molecular assemblies. It was found that as molecules become more polarised, they become less polarisable. Charge flow between two water molecules is only significant along the direction of the hydrogen bond.

1. Introduction

The molecular polarisability is important for understanding various optical and electrical properties. Standard *ab initio* packages such as Gaussian [1] can be used to calculate this polarisability, but these packages do not provide information about the distribution of this response over the molecule. This becomes significantly important for elongated molecules such as surfactants and polymers.
A general theory on distributing response over a molecule was developed by Stone [2]. Subsequently, Ángyán *et al.* developed the Topological Partioning of Electronic Properties (TPEP) approach [3]. This is a robust scheme to calculate distributed polarisabilities under the coupled perturbed Hartree-Fock (CPHF) [4] and the Atoms in Molecules (AIM) partioning technique [5]. Details of the theory and computational method are described elsewhere [6]. In this paper we seek to enhance the understanding of the context in which the work in ref 6, was discussed. In particular, we investigate the change in electronic properties of molecular assemblies as they are allowed to interact. In addition to this the charge flow for the water dimer under external field is calculated.

A. Driessen (ed.), Nonlinear Optics for the Information Society, 107–110.

The level of interaction is described in section 2. This is followed in section 3 by the results for the electronic properties and charge flow of the water dimer. Our conclusions are presented in section 4.

2. Level of interaction

What happens in terms of the electron density as molecules are allowed to interact? And how does that affect the electronic properties? Consider two molecules, such as for example in the water dimer. If these molecules are not allowed to interact with each other, such as is shown in figure 1, then their respective electronic properties such as dipole moment (μ) and polarisability (α) are equivalent. Allowing the molecules to interact, results in a change in electron density, as is shown schematically in figure 1.

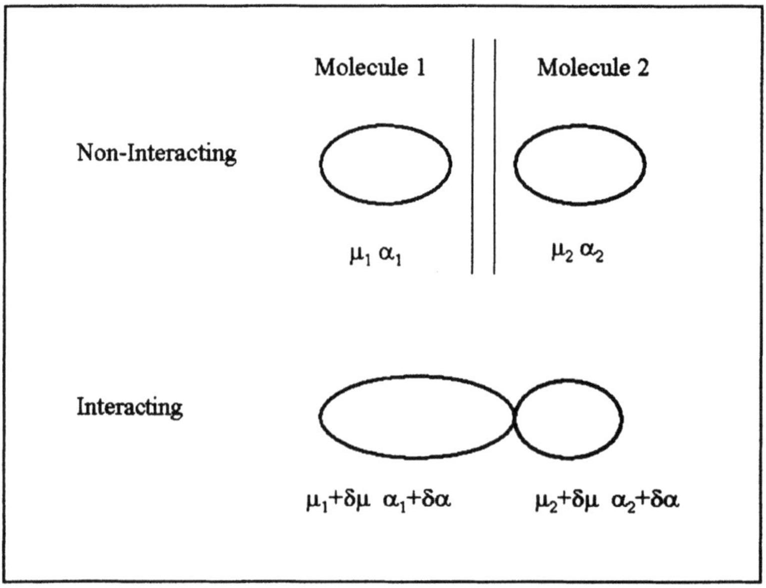

Figure 1. A pictorial representation of the difference in electron density (circles) and electronic properties (μ,α) in molecular assemblies.

This pictorial representation signifies that the change in electron density is not equal for both molecules. In the case of the water dimer this is a result of the formation of the hydrogen bond. Due to this change in electron density the electronic properties will change. This change in electron density results in induced properties ($\delta\mu$, $\delta\alpha$) due to interactions between the molecules in the assembly.

3. Results

The molecular electronic properties have been calculated for a water dimer in the experimental global minimum (linear structure) geometry [7] illustrated in figure 2, using the TPEP approach. The dimer is a simple example of a molecular assembly discussed in section 2. Calculations have been performed for the dimer where both molecules are not allowed to interact, and these are compared with the dimer in which both molecules are allowed to interact.

TABLE 1. Comparison between dipole moments for interacting and non-interacting dimer.

Dimer	Mol 1 / D	Mol 2 / D	Net / D
Non-interacting	1.95	1.95	2.51
Interacting	2.09	2.16	2.97

As the molecules in the dimer are allowed to interact the dipole moments of both molecules increase (see table 1) as a result of reorganisation of electrons due to hydrogen bond formation. This enhancement of dipole moment in the interacting dimer is well known [8].

TABLE 2. Comparison between average molecular polarisability of the non-interacting and interacting dimer

Dimer	Mol 1 / au	Mol 2 / au
Non-interacting	8.3	8.3
Interacting	7.5	7.8

Calculating molecular polarisabilities in molecular assemblies is not possible using standard *ab initio* packages. In the TPEP approach distributed polarisability components[2,3,5] are used to calculate individual molecular polarisabilities.
The effect of interaction in molecular assemblies can be seen from the average molecular polarisabilities, see table 2. The average polarisability is follows from the trace of the polarisability tensor. Owing to hydrogen bond formation both molecules become less polarised, but molecule 1 is more affected than molecule 2.
Thus the results for the water dimer show that as molecules become more polarised, they become less polarisable.
Placing the molecular assembly of the water dimer in an uniform external field showed that charge flows is only significant if the field is along the direction of hydrogen bond.

4. Conclusions

We have shown in the context of hydrogen bonding that as molecules become more polarised they become less polarisable. Placing the dimer in a uniform external field showed that charge flow is only significant along the direction of the hydrogen bond. When an electric field is applied to a molecule several things happen, the properties within an atomic basin change, and the atomic basin itself changes. The calculations that we have presented only take account of the atomic basin change, the atoms are taken to be "frozen". However, the molecular properties obtained in this context are the

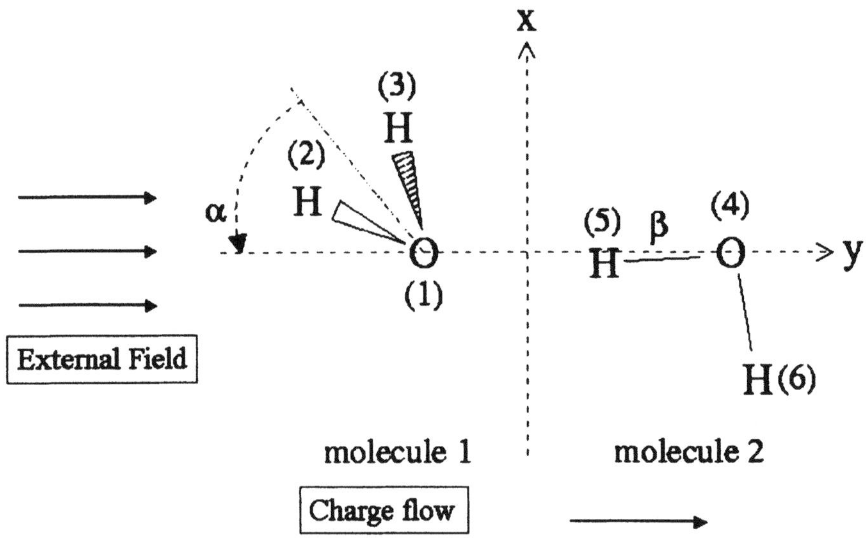

Figure 2. Global minimum geometry of the water dimer.

same, because the changes in atomic basins only affect atomic properties and not molecular properties.

Acknowledgements

We would like to thank the Irish Higher Education Authority (HEA). MihP acknowledges support from the European Union DELOS network programme.

References

[1] Gaussian 94 (Revision A.1), M.J. Frisch, G.W. Trucks, H.B. Schlegel *et al.* (Gaussian Inc., Pittsburgh PA, 1995).
[2] A.J. Stone, *Mol. Phys 56*, 1065(1985)
[3] J.G. Ángyán, G. Jansen, M. Loos, C. Hättig and B.A. Heß, *Chem. Phys. Lett. 219*, 267 (1994).
[4] R. McWeeney, *Methods of molecular quantum mechanics* (Academic Press, New York, 1989).
[5] P.L.A. Popelier, *Atoms in molecules – an introduction* (Pearson Education, Harlow, 2000).
[6] M. in het Panhuis, P.L.A. Popelier, R.W. Munn and J.G. Ángyán, submitted (2000).
[7] C. Millot and A.J. Stone, *Mol. Phys. 77*, 439 (1992).
[8] P.L. Silvisteelli and M. Parinello, *J. Chem. Phys. 112*, 3285 (2000).

POLYMETHINEIMINE CHAINS AS NEW MATERIALS PRESENTING LARGE FIRST HYPERPOLARIZABILITY

D. JACQUEMIN AND B. CHAMPAGNE
Laboratoire de Chimie Théorique Appliquée,Facultés Universitaires Notre-Dame de la Paix,
61 rue deBruxelles, B-5000 Namur (Belgium)
Tel : 00 32 81 724557, Fax : 00 32 81 724567,djacquem@scf.fundp.ac.be

Abstract

The longitudinal first hyperpolarizability (β) of polymethineimine oligomers of increasing size has been evaluated by using a wide panel of *ab initio* and semi-empirical methods. The specific evolution with chain length of β is analyzed in terms of delocalization and asymmetry, as well as in the framework of the N-state analysis. The effects upon β of the chain conformation and of the substitution of the chain ends by donor/acceptor groups have been addressed. It turns out that both the electronic and vibrational contributions to β of the PMI polymer can be much larger than for classical push-pull systems showing a need for experimental studies on these attractive materials.

Full article not available

A. Driessen (ed.), Nonlinear Optics for the Information Society, 111.

NUMERICAL ANALYSIS OF OPTICAL NONLINEARITIES IN MULTISPAN DWDM FIBRE TRANSMISSION SYSTEMS

M. MARCINIAK, A. SEDLIN
National Institute of Telecommunications, Szachowa 1, 04-894 Warsaw, Poland

Abstract

There are reported results of an investigation of the influence of optical non-linear phenomena: Four Wave Mixing, Stimulated Raman Scattering, Self Phase Modulation, and Cross Phase Modulation on transmission quality in multispan DWDM systems. For the purpose of the analysis the Non-linear Schrödinger Equation description of the optical signal propagation in a non-linear fibre is applied and the equation is resolved numerically both in time and frequency domain. Considered transmission parameters include extreme cases up to 128 WDM channels, 200, 100, and 50 GHz channel spacing, 7 transmission spans each 120 km long. New types of transmission fibres are considered. Guidelines for an optimised design of new transmission systems or for existing system upgrades are reported.

1. Introduction

Optical fibre telecommunication networks develop extremely fast in recent years. The newly introduced wideband services as "video on demand" and videoconferences or emerging Internet services force telecommunication operators to increase the capacity of the network infrastructure. Existing transmission links are progressively upgraded mainly via introduction of new techniques as dense wavelength-division multiplexed (DWDM) transmission technology.

Modern high-speed DWDM systems are typically built of several transmission spans, each consisting of an erbium-doped fibre amplifier, a single-mode fibre transmission section, and the dispersion compensation section (typically a piece of dispersion compensating fibre or a chirped Fibre-Bragg Grating). Such cascaded configuration leads to an accumulation of the products of non-linear optical interactions. That in turn results in an increase of the optical interchannel crosstalk and degrades the temporal and spectral characteristics of the signal, including the decrease of signal to noise ratio. Consequently, strong limitations for the number of channels, the channel spacing, the bit-rate and transmission distance occur.

Major problem of the network upgrade is to know to what extent the already existing infrastructure can be modernised. As a consequence, the network designers should be

A. Driessen (ed.), Nonlinear Optics for the Information Society, 113–121.

aware of the limitations for number of channels, maximum transmission speed, as well as the distance between neighbouring optical amplifiers. Those system parameters are determined by fibre attenuation, dispersion, and the optical noise level which results from the non-linear optical phenomena in the silica fibre itself.

The non-linear optical interactions are an essential factor for DWDM systems designers. Usually, the stimulated Raman scattering is stronger for wider channel spacing. On the other hand, narrow channel spacing results in an increase of Cross Phase Modulation and Four-Wave Mixing. Therefore it is essential to know which phenomenon is stronger and has greater influence on a particular system of interest.

The simplest rule is the more channels with higher bit rate, the shorter distance between amplifiers (or regenerators). Dispersion in fibres limits the distance of transmission exponentially with the bit rate, but this problem could be easily overcome with the use of dispersion compensation fibres (DCF) combined with standard single-mode fibres (SMF). Unfortunately, common DCFs have a high attenuation coefficient and decrease the optical power in transmission line so additional amplifiers or higher gain are needed to compensate for the losses.

Moreover, the transmission system working on higher average optical power is more susceptible for non-linear distortions of the signal. In multichannel systems the total power is the sum of power in particular channels. Signals that co-propagate in the neighbouring channels interact particularly strongly and produce unpredictable noise and also a decrease of the signal-to-noise ratio (SNR) in those channels. Those phenomena are to be carefully investigated, especially in the case of utilising new fibre types with decreased dispersion as Lucent Technologies TrueWave RS, Corning LEAF, and Alcatel TeraLight.

In this report we discuss the influence of optical non-linear phenomena: Four Wave Mixing (FWM), Stimulated Raman Scattering (SRS), Self Phase Modulation (SPM), and Cross Phase Modulation (XPM) on transmission quality in multispan DWDM systems. The detailed description of the modelling of non-linear phenomena in silica fibres is not provided here and it can be found e.g. in [1][2]. For the purpose of the analysis the Non-linear Schrödinger Equation description of the optical pulse propagation in the fibre is applied and consequently the equation is resolved numerically both in time and frequency domain.

The considered transmission parameters include case with bit rate of 10 Gbit/s. Number of WDM channels is from 16 to 128. The optical carrier frequency channel spacing is set to 200, 100, 50 and 25 GHz. We restrict ourselves to the C (conventional) transmission band, i.e. 1530 – 1570 nm. Under that limit the maximal channel spacing is inverse-proportionally dependent on the number of channels, e.g. 100 GHz for 32 channels, 50 GHz for 64 channels and 25 GHz for system with 128 channels. The number of transmission spans is up to 7, and single span length is up to 120 km. We include systems with new types of transmission fibres as Corning SMF-28, Lucent Technologies TrueWave RS, Corning LEAF, and Alcatel TeraLight.

Finally we propose guidelines for an optimised design for existing system upgrades, or for new transmission systems using commercially available optical fibre types.

2. Multispan system modelling

A model system to be analysed is shown in Figure 1. It consists generally of N transmission sections (N is equal up to 7). Each section consists of a single mode fibre, an optical amplifier, and (optionally) a dispersion compensation fibre, which length is chosen to compensate optimally for the signal dispersion in the SMF fibre. It is assumed that the optical power in DCF is low enough the non-linear effects be negligible in it. The second optical amplifier after the DCF fibre has an optical gain equal to the absolute value of losses in DCF. Total length of multispan system does not exceed 420 km and this limitation is typical for most European systems.

The most efficient way to minimise intermodulation between channels (Cross Phase Modulation) is to maintain power differences as small as possible. Numerical analyses and also laboratory tests done by optical equipment manufacturers have proven the necessity of the channel power equalisation in long distance systems with 80 or more channels. Therefore it is necessary to install optical amplifiers equipped with automated gain control system in order to assure that the power deviations do not exceed 2-3 dB.

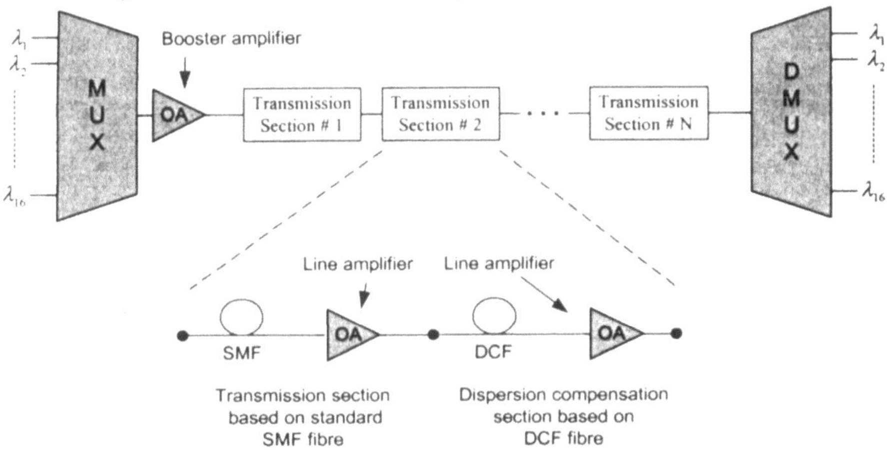

Figure 1. Configuration of multispan DWDM system analysed in numerical simulation.

The important issue is concerned with the limitation of power from booster amplifiers. International standards limit the aggregate signal power in the fibre to 17 dBm. That value gives sufficient single-channel level for few (4, 8, 16) channels, but it leads to a fraction of dBm per channel in larger systems with 64 or 128 channels, so this factor becomes critical. Note that lower power for each channel means the signal less resistant to the noise. Fortunately, the results reported here show the possibility of increasing the 17-dBm threshold up to 19 dBm, or even higher in short distance multichannel fibre links.

Parameters of considered fibre types are listed in Table 1.

A set of simulation models for single and multispan transmission systems with different fibre types listed in Table 1 was developed for 10 Gbit/s transmission speed, varying number of channels (16, 32, 40, 64, 80, 128), and for several values of inter-channel frequency spacing (200, 100, 50 and 25 GHz). Simulations were performed for various aggregate power levels. The critical value of the Bit-Error Rate (BER) coefficient for satisfactory performance of the system was set to be at least 10^{-12}.

TABLE 1. Transmission characteristics of most popular fibre types.

	Manufacturer and type of fibre	Chromatic dispersion $\lambda \cong 1550nm$	Dispersion slope	Effective core area (A_{eff})	Non-linear coefficient (γ)
		Ps/nm·km	Ps/nm^2·km	μm^2	1/W·km
A	Corning SMF-28 (G.652)	+16	0,080	80	1,5
B	Lucent Tech. TrueWave RS (G.655)	+3,0	0,045	55	2,2
C	Corning LEAF (G.655)	+3,0	0,090	72	1,7
D	Alcatel TeraLight (G.655)	+10,0	0,080	65	1,8

The main objective of this investigation is to understand how the non-linear products accumulate in medium distance (up to 420 km) fibre transmission systems. Another goal is to evaluate the usefulness of different kinds of fibres to DWDM systems. The choice of fibre is dependent on its transmission features like maximal power level that can be transmitted through fibre link while keeping non-linear distortions sufficiently low (it strictly depends on the non-linearity coefficient of the particular fibre type), dispersion coefficient and attenuation, which factors set the span length limitation.

3. Theory and notation

The algorithm used in the analysis solves the non-linear Schrödinger equation with the use of the split-step Fourier method. The model takes into account stimulated Raman scattering (SRS), four-wave mixing (FWM), self-phase modulation (SPM), and cross-phase modulation (XPM) phenomena. The first order group-velocity dispersion (GVD), second order GVD and attenuation of the fibre are the input data of the calculations. It should be noted however, that the algorithm does not support the stimulated Brillouin scattering, so we do not discuss this phenomenon in the present work. We do not believe that the results would be significantly altered by SRS, as this phenomenon requires phase-matching condition[3] what is hard to fulfil in because of dispersion of the fibre.

SCHRÖDINGER EQUATION

The generalised non-linear Schrödinger equation description of the propagation of optical signals in non-linear fibres is given by

$$\frac{\partial \underline{E}(z,t)}{\partial z} = [\underline{\hat{D}} + \underline{\hat{N}}] \cdot \underline{E}(z,t) \tag{1}$$

where $\underline{E}(z,t)$ denotes the electric field of the light wave, $\left|\underline{E}(z,t)\right|^2$ characterises its power, $\hat{\underline{N}}$ is the non-linearity operator, and $\hat{\underline{D}}$ is the dispersion operator that is given by

$$\hat{\underline{D}} = j\frac{\beta_2}{2}\frac{\partial^2}{\partial t^2} + \frac{\beta_3}{6}\frac{\partial^3}{\partial t^3} - \frac{\alpha}{2} \tag{2}$$

Here β_2 [s^2/m] describes the first order group-velocity dispersion (GVD), β_3 [s^3/m] is the second order GVD and α[1/m] is the attenuation constant. The attenuation parameter a [dB/m] is related to α according to $a=\alpha\cdot 10 \log e$. First and second order GVD, β_2 and β_3, can be expressed in terms of the dispersion coefficient D_λ and its slope $S_\lambda=dD_\lambda/d\lambda$:

$$\beta_2 = -\frac{\lambda^2}{2\pi c}D_\lambda \tag{3}$$

$$\beta_3 = \frac{\lambda^2}{(2\pi c)^2}(\lambda^2 S_\lambda + 2\lambda D_\lambda) \tag{4}$$

STIMULATED RAMAN SCATTERING

The exclusion of the stimulated Raman scattering from the analytical model leads to a simplified form of the non-linear operator $\hat{\underline{N}}$

$$\hat{\underline{N}} = -j\gamma\left|\underline{E}(z,t)\right|^2 \qquad \text{with} \qquad \gamma = \frac{2\pi n_2 f_{ref}}{cA_{eff}} \tag{5}$$

The operator depends on the non-linear coefficient n_2, effective core area of fibre A_{eff}, as well as on the reference frequency f_{ref}.

In contrary, when stimulated Raman scattering is included in the model then the non-linear operator $\hat{\underline{N}}$ is extended to

$$\hat{\underline{N}} = -j\gamma(1-f_r)\left|\underline{E}(z,t)\right|^2 - j\gamma f_r\int_0^\infty h_r(\tau)\left|\underline{E}(z,t)\right|^2 d\tau \tag{6}$$

where f_r represents the fractional contribution of the delayed Raman response $h_r(t)$ which is approximated by

$$h_r(t) = \frac{\tau_1^2+\tau_2^2}{\tau_1\tau_2^2}\exp\left(-\frac{t}{\tau_2}\right)\sin\left(\frac{t}{\tau_1}\right),\ t>0 \tag{7}$$

In this equation τ_1 and τ_2 are two adjustable parameters to provide a good fit to the Raman-gain spectrum.

For parameterised signals and noise bins the effect of stimulated Raman scattering can be described by the following set of ordinary differential equations

$$\frac{dA_i}{dz} = -\frac{\alpha_i}{2} A_i - jf_r\gamma \sum_{m \neq i} A_i |A_m|^2 j \operatorname{Im}(\widetilde{R}(f_i - f_m)) \tag{8}$$

$$\frac{dN_k}{dz} = -\frac{\alpha_k}{2} N_k - jf_r\gamma \sum_{i,k} N_k |A_i|^2 j \operatorname{Im}(\widetilde{R}(f_k - f_i)) \tag{9}$$

where A_i is the amplitude of the ith parameterised signal, N_k is the amplitude of the kth noise bin, and α_i, α_k represent their attenuation coefficients, respectively, and f_r describes the fractional contribution of the delayed Raman response to the non-linearity coefficient γ, and Im($\widetilde{R}$ $(f_i$-$f_m)$) represents the imaginary part of the Fourier transformed Raman response function at the frequency difference between the parameterised signals $(f_i$-$f_m)$. In the second equation, Im($\widetilde{R}$ $(f_i$-$f_m)$) denotes the imaginary part of the Fourier transformed Raman response function at the frequency difference between the noise bins and the parameterised signals. The above set of equations can to be solved using a 4th order Runge-Kutta equation solver.

4. Numerical results

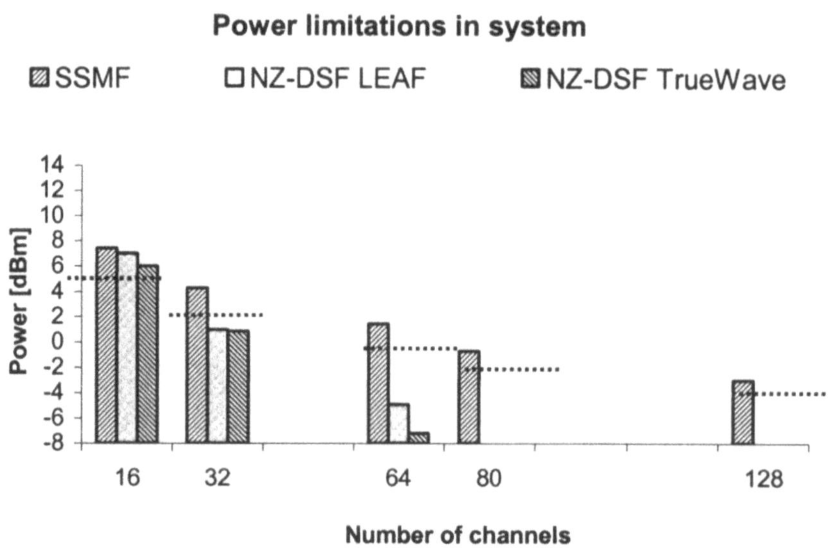

Figure 2. Maximal optical power possible to propagate through various fibre types in multi- span WDM systems: 16/200 (number of channels / channel spacing), 32/100, 64/50, 128/25. The dashed lines represent power limitation from 17 dBm standard threshold.

Figure 2 shows the maximal optical power acceptable for several fibre types in multi-span WDM systems with parameters (number of channels / channel spacing) as follows: 16/200 32/100, 64/50, and 128/25. It can be seen on the figure how high could be power in single channel without exceeding the non-linear limitation (which was assumed as 4 dB decreasing of SNR). The power decreases with increasing number of channels, however, at rate varying with fibre type. The best is the standard single mode fibre G.652. Other fibres, e.g. NZ-DSF fibres, have higher non-linear coefficient, what results in faster accumulation of non-linear products along the transmission line. The results indicate also the possibility to increase the 17 dBm power limit recommended by ITU-T international standards as a maximal safe power in WDM systems by additional 2 dB, up to 19 dBm.

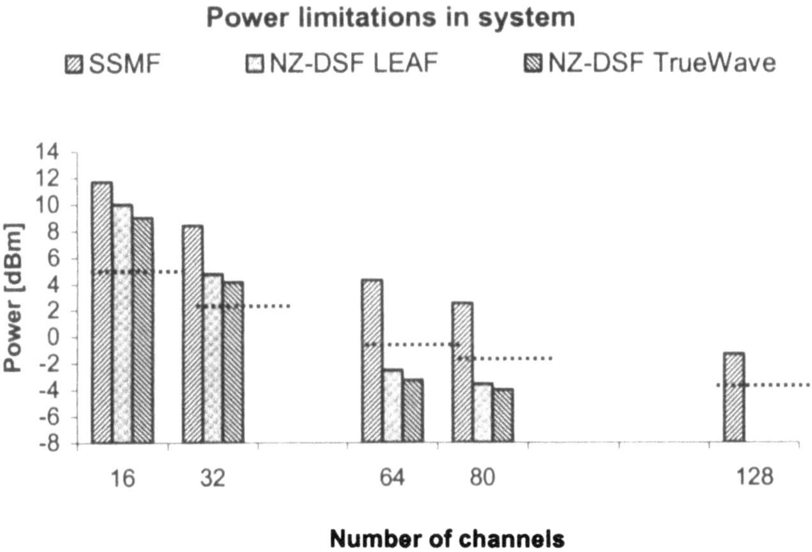

Figure 3. Maximal optical power possible to propagate along various fibre types in single-span WDM systems for 16, 32, 64, and 128 channels. The dashed lines represent 17 dBm power limitation.

Figure 3 displays the maximal optical power in 7 x 60 km length fibre line in different WDM systems. The diagram proves the superiority of G.652 standard single mode fibre over the NZ-DSF fibre, especially when the number of channels is high. In NZ-DSF fibres the main problem is with strong FWM phenomenon, so the channel number, possible to realise in these systems is less comparing to G.652 fibres. In single-span DWDM system, the 17 dBm threshold could be also increased up to 22-23 dBm.

COMPARISON OF SYSTEMS WITH 1 AND 7 AMPLIFIER SECTIONS

The maximal power acceptable in a 40 channel systems is shown in Figure 4 both for single- and multispan WDM systems, with channel spacing 100 and 50 GHz. The distances are 60 and 420 km. Different maximal power between multispan systems with 100 and 50 GHz channel spacing (right diagram) proves the possibility of a significant increase of the system performance by decreasing the SRS phenomenon influence. Note that it is useless to decrease channel spacing for a short-distance single-span transmission line, what is also very interesting.

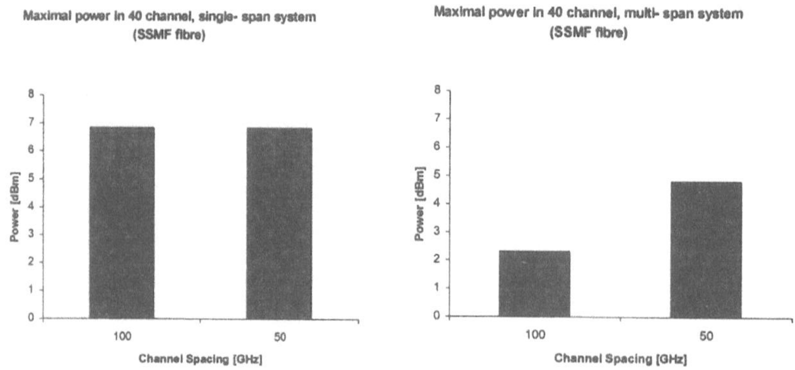

Figure 4. Maximal power possible to achieve in 40 channel single span (on the left) and multispan (on the right) WDM systems with channel spacing 100 and 50 GHz.

5. Conclusions

The results of the multi-span ITU-T DWDM network modelling presented in the paper prove a possibility of increasing 17 dBm safety threshold up to at least 19 dBm in multi-span systems. The threshold could be increased further up to 22-23 dBm in single-span DWDM system. The results also indicate that the conventional standard single mode fibre exhibits a high transmission performance in high-power multi-span DWDM links. In addition to a costly procedure of network modernisation by replacing already existing fibre links by the new fibre types, the standard single mode fibre still remains a competitive choice for high-performance non-linear DWDM transmission network. In order to fight efficiently against the signal-destructive optical non-linearities the use of low non-linearity coefficient fibres should allow construction of transmission links with higher bit-rate, with more channels, and also at longer distances.
The Stimulated Raman Scattering phenomenon is responsible for most signal distortions in comparison with other non-linear phenomena, especially for ITU-T 100 GHz channel frequency spacing. However, when the number of channels becomes very high and channel spacing is as narrow as 25 GHz, then the other non-linear interactions become critical.
Since the non-linear products accumulate in the consequent transmission sections in a very complex way, it is hard to predict correctly the power transfer simply by adding the power from non-linear interactions in next sections. In fact, the accumulation of the noise generated by preceding spans along the subsequent spans is an essential factor. Each channel which power was increased in a particular transmission section starts to transfer its own power to neighbouring channels in the next section by Cross Phase Modulation and Four Wave Mixing, as well as to the channels at lower frequencies by Stimulated Raman Scattering. Simultaneously the particular channel takes also power from other channels. In general the behaviour of each channel is different so it is hard to establish simple design rules for long distance transmission lines.

Acknowledgements

The authors kindly acknowledge their participation in the project "Analysis of the non-linear crosstalk and channel spacing optimisation in DWDM systems in real fibre links" sponsored by the State Committee for Scientific Research in 1999-2000. The PTDS (Photonic Transmission Design Suit) software has been used for a part of the simulations.

References

[1] G. P. Agrawal, *Nonlinear Fiber Optics*, Academic Press, San Diego, (1997).
[2] M. Kowalewski, M. Marciniak, and A. Sedlin, Nonlinear interactions in wavelength-multiplexed optical fiber telecommunication systems, *J. Opt. A: Pure Appl. Opt. Vol. 2*, No. 4, pp. 319-326 (2000).
[3] L. B. Jeunhomme, *Single-Mode Fiber Optics, Marcel Dekker, Inc., New York,* (1990).

NONLINEAR CHARACTERIZATION OF VERY THIN POLYMER FILMS BY SURFACE PLASMON RESONANCE

G. MARGHERI, E. GIORGETTI, F. GELLI, S. SOTTINI
Istituto di Ricerca sulle Onde Elettromagnetiche "Nello Carrara"
Via Panciatichi 64, 50127 FIRENZE, Italy
Tel. +39 055 4235254, Fax. +39 055 410893 sottini@iroe.fi.cnr.it

Abstract

The investigation of new polymers with large third order nonlinearities is of great interest in view of applications to all optical processing. It was early recognized that, depending on the characteristics of the film under investigation different techniques are appropriate for measuring its third order susceptibility $\chi^{(3)}$. In practice the synthesis of a small amount of a novel material is costly and time consuming, as well as its subsequent processing to realize films; as a consequence , at least at an early stage, the nonlinear tests have usually to deal with small quantities of material and with the poor quality of the obtained samples. For these reasons the Surface Plasmon Resonance (SPR) technique is very attractive for the linear and nonlinear characterization of very thin films. Limiting ourselves to the nonlinear characterization, to our knowledge only few tests were reported so far, although the intrinsic drawbacks of SPR (TM polarized light and a metal film are required) are largely counterbalanced by a number of advantages, in particular the capability of monitoring sign, real and imaginary part of the nonlinearity of extremely thin layers with a high degree of insensitivity to their defects. Here we report on the use of SPR to evaluate the $\chi^{(3)}$ of thin polymer films (10-150nm). The potential of this method was investigated by numerical simulations. Experimental tests, which were performed at λ=1065nm, by using thin film samples of a novel polydiacetylene, PDCHD-HS, confirmed such potential provided that also the behaviour of the metal film (typically Ag), at low and high intensity, is taken into account. The tests on PDCHD-HS showed a very large value of the real part of the $\chi^{(3)}$ which turned out to be ~ -1.2×10^{-17} m^2/V^2. If this encouraging result is confirmed by further tests at 1300-1500nm, this material could be an interesting candidate for future applications.

Full article not available

A. Driessen (ed.), Nonlinear Optics for the Information Society, 123.

CAVITY SPATIAL SOLITONS IN SEMICONDUCTOR MICROCAVITIES

M. PERRINI[1], T. MAGGIPINTO[1], F. RIZZI[1], M. BRAMBILLA[1], G. TISSONI[2], L. SPINELLI[2]
1) *INFM, Dipartimento Interateneo di Fisica, Università e Politecnico di Bari, Via Orabona 4, 70126 Bari, Italy*
2) *INFM, Dipartimento di Fisica dell'Università, Via Celoria 16, 20133 Milano, Italy*

Keywords

cavity spatial soliton, pattern formation, semiconductor microcavities.

Abstract

In the framework of Transverse Nonlinear Optics and Optical Pattern Formation, we study the formation of cavity solitons (CS) in semiconductor microcavities. We adopt a microscopic model that describes the field and carrier dynamics in the quasi-equilibrium regime, within the free-carrier approximation, with the Urbach tail and the bandgap renormalization, taken into account in a phenomenological way. We numerically predict the formation of patterns in the transverse profile of the output field, in specific range of parameters.

1. Introduction

Cavity Solitons (CS) in two transverse dimensions have recently been predicted in broad-area semiconductor heterostructures such as VCSELs and MQW microresonators [1][2], driven by a coherent external field. They appear as independent and self-confined peaks of light embedded in a homogeneous background. At difference from other systems [3][4], semiconductors are the best candidates for applications to information technology, although they present additional features such as carrier diffusion. Here we study the formation of CS in a microresonator with an active layer constituted by bulk GaAs, with pure photoinjection [5][6]. In section 2 we present the dynamical equations of the model, with the description of the nonlinear susceptibility. Section 3 is devoted to the numerical description of the spatial patterns, and stable CS domains. The stability properties of CS are described in section 4. Section 5 contains a comparison with experiments. Conclusions appear in section 6.

A. Driessen (ed.), Nonlinear Optics for the Information Society, 125–130.

2. Model

Following the same derivation as in Ref. 7 we consider here a broad-area vertical-cavity microresonator filled with bulk GaAs at room temperature, and driven by an external coherent field with a homogeneous profile. In the slowly varying envelope approximation and in the mean field limit the dynamical equations reduce to the following set of partial differential equations [7]:

$$\frac{\partial E}{\partial t} = -(1+i\vartheta)E + E_I + i\Sigma\chi_{nl}E + i\nabla_\perp^2 E,$$
$$\frac{\partial N}{\partial t} = -\gamma\left[N - \mathrm{Im}(\chi_{nl})|E|^2 - d\nabla_\perp^2 N\right] \tag{1}$$

where E and E_I are, respectively, the normalised slowly varying amplitudes of the intracavity field and external driving field; θ is the cavity detuning; Σ is the bistability parameter; N is the carrier density scaled to its transparency value; γ is the normalised, nonradiative, carrier decay; d is the diffusion coefficient. The transverse Laplacian describes diffraction in paraxial approximation and χ is the complex nonlinear susceptibility that describes radiation-matter interaction in the active material. In the quasi-equilibrium approximation, the free carrier susceptibility has the form[7]:

$$\chi_{nl}(N,\omega_0) = -\frac{i}{\varepsilon_0 \hbar V_A}\sum_{\vec{k}}|\mu_k|^2 \frac{f_{ek}(N)+f_{hk}(N)-1}{i(\omega_k-\omega_0)+\gamma_p} \tag{2}$$

where $\vec{k}$ is the carrier momentum, μ_k is the dipole matrix element between the valence and the conduction band, V_A is the active volume, $\gamma_p \cong 10^{13}s^{-1}$ is the polarisation decay rate, and $\hbar\omega_k = \varepsilon_{gap} + \hbar^2k^2/2m_R$ is the transition energy at the carrier momentum k, m_R being the electron-heavy hole reduced mass[6]. The $f_{ek,hk}(N)$ are Fermi-Dirac distributions for electrons and holes, respectively. In the regime of interest, two many-body corrections are then phenomenologically introduced to properly describe the behaviour of the bulk medium: one is bandgap renormalization and the other is the Urbach tail, which are due to the carrier-carrier and the carrier-phonon interactions [7].

3. Analysis of the model

By setting the time derivatives equal to zero and neglecting the Laplacian in equations (1), we obtain the stationary homogeneous solutions (E_S, N_S):

$$|E_I|^2 = |E_S|^2 \left\{ [1 + \Sigma \operatorname{Im}(\chi_{nl})]^2 + [\theta - \Sigma \operatorname{Re}(\chi_{nl})]^2 \right\},$$
$$|E_S|^2 = \frac{N_S}{\operatorname{Im}(\chi_{nl})}. \tag{3}$$

A linear stability analysis[7] allowed us to find the instabilities of the homogeneous steady state that can give rise either to another homogeneous solution (plane wave instability PWI) or to a spatially modulated pattern (MI). We are interested in the latter. By analysing the so-called Turing domain, we obtained information about the parameter ranges where CS may exist [8], and we can outline some favourable and/or necessary conditions for the existence of CS, namely:
a bistable steady-state curve for the homogeneous fields;
the stability of the lower intensity branch of the homogeneous curve;
a large interval of values of the injected field for which the upper intensity branch is modulationally unstable.
The nature of the instabilities and the fulfilment of conditions a), b), c), can be appreciated through the analysis of the PWI and MI boundaries.

3.1 NUMERICAL RESULTS

Following the path outlined in Ref. 7 we performed a detailed study of the instabilities in the parameter space. We made some simulations by fixing various values of the band-gap detuning parameter $\Delta = (\omega_{gap} - \omega_0)/\gamma_p$, and scanning the values of Σ and of θ . In Fig.1.a we have an indication that, for $\Delta = 1$, Σ should be between 65 and 100 to observe MI for accessible values of θ , e.g. around –9.
We also have an indication on existence of a threshold for Σ for observation of both MI and PWI. In Fig.1.b we focus onto the case $\Delta = 1$ and , close to the experimental value, we see that for $-10 < \theta < -7.5$ the curve is bistable, the upper branch is modulationally unstable and the lower branch is stable except for a small part, so a good compromise is given by $\theta = -9$. The dynamical equations (1) were integrated numerically by a split-step method with periodic boundary conditions, and we have found roll and honeycomb solutions together with a CS branch. A long streak of simulations assessed control techniques for CS towards all-optical information encoding and processing. We analysed the on-off switching of CS in any desired position of the transverse plane, by adding to the homogenous input field a narrow Gaussian pulse, centred at the point (x_0, y_0) where we want to create or annihilate the CS. [4][7]
Numerical results are reported in Tables 1 and 2.

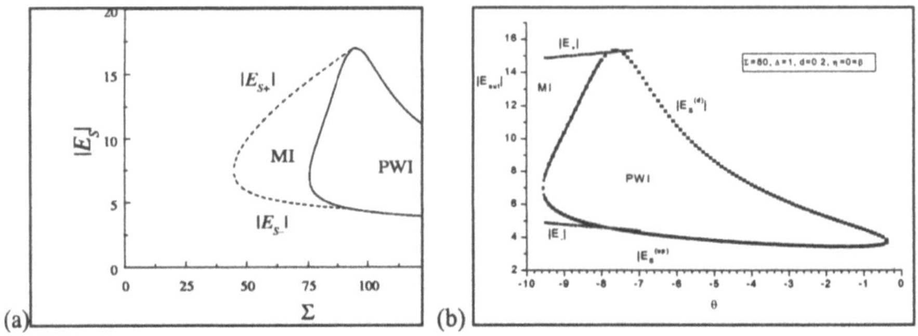

Figure.1.a. MI and PWI boundaries as functions of the parameter Σ. The other parameters are $\Delta = 1, \theta = -9, d = 0.2$. *Figure.1.b.* MI and PWI boundaries as function of the cavity detuning θ.

We have also analysed the interaction properties of CS; for the case under study two solitons are independent if their distance is greater than d=54.4 μm (the width of a soliton is $\approx 20\mu m$) and they merge together in one soliton for smaller peak to peak distances.

TABLE 1. Minimum power needed to excite a CS for different values of the width σ. Pulse duration is T=235*ns*.

Parameter			Value		
$\sigma(\mu m)$	11.76	14.7	17.64	20.58	23.52
$p_{\min}(\mu W)$	98	150	211	288	376

TABLE 2. Minimum power required for erasing an existing CS for different values of the width. *The pulse has a* $\varphi = \pi$ phase shift respect to E_I, and pulse duration of T=23.5 *ns*.

Parameter			Value		
$\sigma(\mu m)$	11.76	14.7	17.64	20.58	23.52
$p_{\min}(\mu W)$	35.8	52	72.4	98.5	124

4. Stationary solutions and stability

Looking for inhomogeneous stationary solutions, $\partial/\partial t = 0$, of equations (1), we can discretised all variable(s) on a spatial grid, using a fast Fourier Transform (FFT) algorithm to numerically evaluate the spatial derivatives[9][10]. This procedure gives a set of coupled nonlinear algebraic equations which can be solved using a Newton method. Given a suitably close initial guess, this method rapidly converges to a stationary solution of the original equations. Once a stationary solution is located its whole branch can then be tracked. The benefit is that this method yields all stationary solutions, not only the stable ones. It also allows to directly evaluate the system's eigenvalues λ describing the solutions' stability and the associated eigenvectors $\vec{u}$. We found 1D CS, and cylindrically symmetric 2D CS. In order to properly describe the latter case we also considered azimuthal perturbations of the form $\varepsilon = R(r)r^{|m|}e^{im\phi}$, where m=0 corresponds to purely radial perturbations. Numerical simulations show that for m=0,1 there are no unstable modes, any instability being related to the $m = \pm 2$ eigenmode. An excellent agreement with dynamical simulation has been found. A complete characterisation of existence, stability and dynamical properties of CS, can be found in Ref. 9.

5. Comparison with the experiment

In an experiment, the profile of the input field can't be a pure PW, thus a Gaussian profile for E_I was introduced and we evaluated its effects on CS behaviour and stability. This field profile causes the local onset of patterns even though the maximum beam intensity is beyond the upper instability threshold. In Fig.3.a we see a pattern in the outer region, where the input field is inside the MI domain, and its modulation is roughly equal to the simulated global structures; while in the centre a non-modulated region approximates the analytical homogeneous solution. Fig.3.b shows the field intensity profile where the structures have the typical shape of the patterns obtained with PW input, but with circular symmetry. Moreover decreasing the incident intensity we should observe the movement of the structures towards the centre, joining each other, and eliminating the central homogeneous part; increasing the incident intensity we should observe just the opposite. These evidences agree with observations [11]. We have also studied the role of external perturbations leading to modifications in the character of pattern-forming instabilities. In the experiment, patterns are met at input field powers much lower than what theoretical predictions suggested. An indication on the possible origin of this phenomenon was found in the role of the spatially distributed irregularities due to the layer jumps, an intrinsic feature of MBE and MOVPE growth processes. These irregularities caused fluctuations of the microresonator resonances throughout the transverse section of the epitaxial sample that can be represented by introducing spatially distributed fluctuations of the cavity detuning θ profile. We have introduced a proper modelling thereof and we gained a clear evidence of a radical lowering of the MI threshold [12].

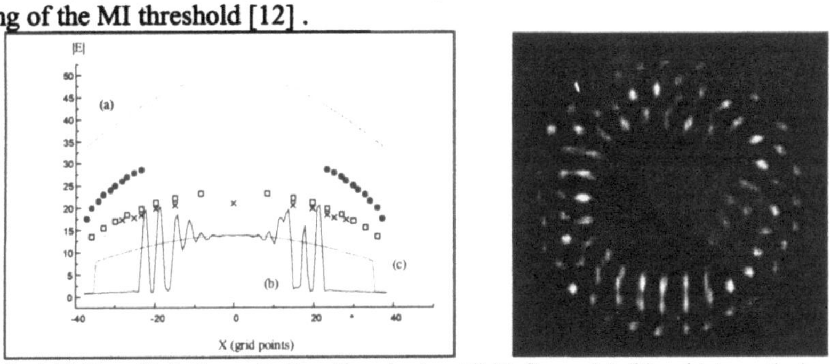

Figure. 3: a). Field profile in the cavity for a Gaussian input field. Curves (a), (b) and (c) are: the input field $|E_{in}|$, the cavity field $|E_{out}|$, the homogeneous stationary solution $|E_S|$. Crosses, squares and circles indicate: honeycombs, rolls and CS. b) Field profile in the cavity for a Gaussian input field.

6. Conclusion

We have presented briefly the main results concerning CS formation in GaAs bulk microresonators highlighting the best operational regimes and parametric choices by which stable CS can be better exploited; we have done the analysis of stability domains for such structures, and we have set forth criteria to predict soliton stability, dynamics and interaction schemes. The reader is referred to bibliography for more detailed information.

References

[1] M. Brambilla, L.A. Lugiato, F. Prati, L. Spinelli and W.J. Firth, *Phys. Rev. Lett. 79*, 2042 (1997).
[2] W.J. Firth and A.C. Scroggie, *Phys. Rev. Lett. 34*, 1623 (1996).
[3] M. Brambilla, L.A. Lugiato and M. Stefani, *Europhys. Lett. 34*, 109 (1996).
[4] L. Spinelli, G. Tissoni, M. Brambilla, F. Prati and L.A. Lugiato, *Phys. Rev. A 58*, 2542 (1998).
[5] H. Haug and S.W. Koch, *Phys. Rev. A 39*, 1887 (1989).
[6] W. W. Chow, S. W. Koch and M. Sargent III, *Semiconductor-Laser Physics*, Springer Verlag, Berlin Heidelberg (1994).
[7] G. Tissoni, L. Spinelli, M. Brambilla, T. Maggipinto, I. M. Perrini, and L. A. Lugiato, *J. Opt. Soc. Am. B 16*, pp. 2083-2094, and following paper pp. 2095-2105, (1999) .
[8] L. A. Lugiato, L. Spinelli, G. Tissoni, and M. Brambilla, *J. Opt.B 1*, 43-51 (1999).
[9] T. Maggipinto, M. Brambilla, G. K. Harkness, and W. J. Firth, Cavity Solitons in semiconductor microresonator: existence, stability and dynamical properties, *accepted to appear in Phys. Rev. E.*
[10] W. J. Firth and G. K. Harkness, *Asian Journal of Physics 7*, pp.665-676, (1998).
[11] G. Slekys, I. Ganne, I. Sagnes, R. Kuszelewicz, *J. Opt. B, vol.2*, n.3, (2000).
[12] M. Brambilla and F. Rizzi, *in preparation*.

ADAPTIVE WAVEGUIDES USING PHOTOREFRACTIVE CREENING SOLITONS

J. PETTER, C WEILNAU, C. DENZ
Institute of Applied Physics, Darmstadt, University of Technology Hochschulstr. 6, D - 64289 Darmstadt, Germany Tel: ++49 6151 163122, Fax: ++49 6151 163122, juergen.petter@physik.tu-darmstadt.de

Abstract

Photorefractive screening solitons appear when self-focusing inside a photorefractive material exactly balances the diffraction of the propagating light beam. The refractive index change induced by these solitons can effectively guide other beams and therefore spatial photorefractive solitons can be used as adaptive waveguides. In time resolved measurements we show that a HeNe-probe beam exactly follows the trajectory of the inducing beam. If two incoherent solitons are launched simultaneously, strong interaction effects (attraction, repulsion, rotation, energy exchange) occur due to their common anisotropic potential. These effects offer the possibility of using photorefractive solitons for the realization of different waveguide devices such as switches, dividers or couplers. For this purpose an exact knowledge of the soliton trajectories during propagation is indispensable. We realized a time- and spatially resolved examination of the interacting solitons, which gives a detailed insight into the trajectories of the beams. Furthermore we present the successful guidance of a probe beam in interacting photorefractive solitons with and without mutual exchange of energy and therefore the realization of an adaptive all-optical switch and a 1-to-2-coupler using photorefractive solitons.

Full article not available

A. Driessen (ed.), Nonlinear Optics for the Information Society, 131.

VIBRATIONAL FIRST AND SECOND HYPERPOLARIZABILITY OF H-C≡N

O. QUINET[1], B CHAMPAGNE[1,2]
1) *Laboratoire de Chimie Théorique Appliquée, Facultés Universitaires Notre-Dame de la Paix, rue de Bruxelles, 61, B-5000 Namur, Belgium.*
2) *Research Associate of the National Fund for Scientific Research (Belgium).*

Abstract

The vibrational first and second hyperpolarizability of hydrogen cyanide has been computed *ab initio* at the RHF and MP2 levels with inclusion of the first-order anharmonicity contributions for the most common NLO processes.

1. Introduction and Computational Procedure

The deduction of structure-property relationships is of central importance for understanding and interpreting the nonlinear optical (NLO) phenomena [1]. This work, part of a general study on the characterization of the vibrational hyperpolarizabilities of small molecules [2], is aimed at determining the vibrational first (β^v) and second (γ^v) hyperpolarizabilities of hydrogen cyanide, H-C≡N by using *ab initio* quantum chemical calculations including electron correlation. Whereas field-induced modifications of the electron density are fast and associated with the electronic (hyper)polarizabilities (α^e, β^e, and γ^e), the changes in equilibrium nuclear positions are one to three orders of magnitude slower and constitute the nuclear relaxation (NR) contribution, *i.e.*, α^{nr}, β^{nr}, and γ^{nr} which often constitutes the main part to the pure vibrational (hyper)polarizabilities, α^v, β^v, and γ^v. The other vibrational contributions originate from the variations in zero-point vibrational energy and are therefore called the curvature contributions, α^c, β^c, and γ^c. In the clamped nucleus approximation, the external electric fields act sequentially rather than simultaneously upon the electronic and nuclear motions and the total (hyper)polarizabilities can be expressed as the sums of the electronic contribution, the ZPVA contribution and the pure vibrational contribution,

$$P = P^e + \Delta P^{zpva} + P^v \quad \text{with } P = \beta, \gamma \tag{1}$$

From the exact summation-over-states (SOS) expressions, Bishop and Kirtman [3] have derived the vibrational hyperpolarizability expressions and have worked out compact expressions under nonresonant conditions as summations of harmonic and anharmonic

A. Driessen (ed.), Nonlinear Optics for the Information Society, 133–137.

terms. In this study we have only considered the lowest-order non-vanishing terms of each type, *i.e.*

$$\beta^{v}_{\zeta\eta\xi} = [\mu\alpha]^{0}_{\zeta\eta\xi} + [\mu^{3}]^{I}_{\zeta\eta\xi} \tag{2}$$

$$\gamma^{v}_{\zeta\eta\xi\chi} = [\alpha^{2}]^{0}_{\zeta\eta\xi\chi} + [\mu\beta]^{0}_{\zeta\eta\xi\chi} + [\mu^{2}\alpha]^{I}_{\zeta\eta\xi\chi} \tag{3}$$

where for instance, $[\mu^2\alpha]^I = [\mu^2\alpha]^{1,0} + [\mu^2\alpha]^{0,1}$. ζ, η, ξ, and χ are associated with the Cartesian coordinates. The notation $[X]^{n,m}$ means n^{th} order electrical anharmonicity and m^{th} order mechanical anharmonicity. The interested reader can find the corresponding summation-over-modes (SOM) expressions in Ref. 3. These expressions can be evaluated for any combination of frequencies as well as for any NLO processes. Within the infinite optical frequency (or enhanced) approximation [4], one assumes that the vibrational frequencies are at least one order of magnitude smaller than the optical frequencies or, in other words that the dispersion of the dynamic pure vibrational hyperpolarizability for a given process is small in the optical wavelength domain. Within this approximation, for the diagonal components as well as for the parallel or isotropic average quantity,

$$\gamma^{K} = 3/2(\gamma_{//} - \gamma_{\perp})$$

$$\gamma_{//} = \bar{\gamma} = \frac{1}{15}\sum_{\xi}\sum_{\eta}(\gamma_{\xi\xi\eta\eta} + \gamma_{\xi\eta\xi\eta} + \gamma_{\xi\eta\eta\xi}) \, and \, \gamma_{\perp} = \frac{1}{15}\sum_{\xi}\sum_{\eta}(2\gamma_{\xi\eta\eta\xi} - \gamma_{\xi\xi\eta\eta})$$

and

$$\gamma^{K} = 3/2(\gamma_{//} - \gamma_{\perp})$$

$$\gamma_{//} = \bar{\gamma} = \frac{1}{15}\sum_{\xi}\sum_{\eta}(\gamma_{\xi\xi\eta\eta} + \gamma_{\xi\eta\xi\eta} + \gamma_{\xi\eta\eta\xi}) \, and \, \gamma_{\perp} = \frac{1}{15}\sum_{\xi}\sum_{\eta}(2\gamma_{\xi\eta\eta\xi} - \gamma_{\xi\xi\eta\eta})$$

(Einstein convention used; z is the direction of the dipole moment), the vibrational contributions to the various NLO processes are expressed as simple fractional multiples of the corresponding quantities in the static limit :

dc-Pockels : $$\gamma^{v}_{L}(-\omega;\omega,0,0)_{\omega\to\infty} = \frac{1}{3}[\alpha^{2}]^{0,0}_{L;\omega=0} + \frac{1}{2}[\mu\beta]^{0,0}_{L;\omega=0} \tag{4}$$

dc-Kerr : $$\gamma^{v}_{L}(-\omega;\omega,0,0)_{\omega\to\infty} = \frac{1}{3}[\alpha^{2}]^{0,0}_{L;\omega=0} + \frac{1}{2}[\mu\beta]^{0,0}_{L;\omega=0} \tag{5}$$

ESHG : $$\gamma^{v}_{L}(-2\omega;\omega,\omega,0)_{\omega\to\infty} = \frac{1}{4}[\mu\beta]^{0,0}_{L;\omega=0} \tag{6}$$

IDRI : $$\gamma^{v}_{L}(-\omega;\omega,\omega,-\omega)_{\omega\to\infty} = \frac{2}{3}[\alpha^{2}]^{0,0}_{L;\omega=0} \tag{7}$$

Within this approximation the vibrational contribution to SHG and THG vanishes. Several investigations [2,5] have concluded that the infinite optical frequency approximation is satisfactory for the dynamic β^v and γ^v whereas in the static limit, higher-order terms can be substantial.
All the calculations have been performed by using the GAUSSIAN94 series of programs [6] at the Hartree-Fock and second-order Møller-Plesset (MP2) perturbation theory levels with the aug-cc-pvdz and aug-cc-pvtz basis sets [7]. The description of the computational and numerical procedures to compute the vibrational hyperpolarizabilities as well as their static electronic counterpart can be found in recent investigations [2].

2. Results and discussion

The electronic and vibrational hyperpolarizability contributions of HCN calculated with the aug-cc-pvtz basis set are listed in TABLE I. Some components of β depend strongly upon the inclusion of electron correlation as well as upon the extend of the basis set. In particular $\beta^e_{//}(0;0,0)$ for which we prefer to use the converged (with respect to both basis set and electron correlation) CCSD(T) value of –2.8 a.u. due to Maroulis and Pouchan [8]. $[\mu\alpha]^0_{//;\omega=0}$ and to a lesser extend the first-order anharmonic contributions are also very sensitive to electron correlation. On the other hand, these electron correlation and basis set effects are reduced for the second hyperpolarizability. The largest effect appears for the first-order anharmonic contributions but, since they go in opposite direction, the global effect upon $[\mu^2\alpha]^I_{//;\omega=0}$ is small. Moreover the CCSD(T) value of Ref. 8 amounts to 220 x 10 a.u., i.e. less than 20% larger than our MP2/aug-cc-pvtz result.
Table 1 also displays the static and infinite optical frequency first and second hyperpolarizabilities for the most common NLO processes. Using the MP2 value as reference, the vibrational contribution to the dc-Pockels first hyperpolarizability is substantial. Using a typical wavelength of 532.8 nm ($\hbar\omega$ = 2.33 eV), $\beta^v_{//}(-\omega;\omega,0)$ and $\beta^{v,K}(-\omega;\omega,0)$ amount to –0.7 a.u. and –4.3 a.u. whereas their infinite optical frequency estimates are 16% and 5% larger. For λ = 694.3 nm and 1064 nm, the vibrational contribution to $\beta_{//}(-2\omega;\omega,\omega)$ attain –0.3 a.u. and –0.7 a.u., respectively. Contrary to β, the vibrational contribution to the second hyperpolarizability is smaller than its electronic counterpart. In the infinite optical frequency approximation, $\gamma^v_{//}(-2\omega;\omega,\omega,0)$ amounts to 3% of $\gamma^e_{//}(0;0,0,0)$ whereas for $\gamma^{v,K}(-\omega;\omega,0,0)$ and $\gamma^v_{//}(-\omega;\omega,-\omega,\omega)_{\omega\to\infty}$ the vibrational component attains 12% and 9%, respectively. Like for acetylene [2], $[\mu^2\alpha]^{1,0}_{//;\omega=0}$ dominates the static vibrational response. At a wavelength of 532.8nm, $\gamma^{v,K}(-\omega;\omega,0,0)$ = 244 a.u. while $\gamma^v_{//}(-2\omega;\omega,\omega,0)$ = 57

a.u. and 200 a.u. for λ = 694.3 nm and 1064 nm, respectively. These values for typical frequencies are in good agreement with the infinite optical frequency results.

3. Conclusions and Perspectives

By adopting a quantum chemistry scheme including electron correlation and using an extended basis set, we have determined the vibrational first and second hyperpolarizabilities of HCN up to first order in electrical and mechanical anharmonicity. On the one hand, we predict that for the dc-Pockels experiment, the vibrational contribution is relatively important although the total response is small. On the other hand, we predict a value of 185 a.u. for $\gamma^{v,K}(-\omega;\omega,0,0)$ - $\gamma^{v}_{//}(-2\omega;\omega,\omega,0)$ whereas the static electronic counterpart amounts to 1799 a.u. . It will be interesting to carry out the corresponding experiments in order to complete this theoretical investigation.

Acknowledgements

O.Q. thanks the "Fonds pour la Formation et la Recherche dans l'Industrie et dans l'Agriculture" (FRIA) for financial support. B.C. thanks the Belgian National Fund for Scientific Research for his Research Associate position. The calculations have been performed on the IBM SP2 of the Namur Scientific Computing Facility (Namur-SCF). The authors gratefully acknowledge the financial support of the FNRS-FRFC, the 'Loterie Nationale' for the convention N° 2.4519.97.

References

[1] B. Champagne and B. Kirtman, in *Handbook of Advanced Electronic and Photonic Materials and Devices*, edited by H.S. Nalwa (Academic Press, New York, 2000), Vol. 9, Chap. 2, 63-126.

[2] O. Quinet and B. Champagne, *J. Chem. Phys. 109*, 10594 (1998); O. Quinet and B. Champagne, *Int. J. Quantum Chem. 80*, 871 (2000)

[3] D.M. Bishop and B. Kirtman, *J. Chem. Phys. 95*, 2646 (1991).

[4] D.M. Bishop, M. Hasan, and B. Kirtman, *J. Chem. Phys. 103*, 4157 (1995); J.M. Luis, J. Martí, M. Duran, J.L. Andrés, and B. Kirtman, *J. Chem. Phys. 108*, 4123 (1998).

[5] D.M. Bishop and E.K. Dalskov, *J. Chem. Phys. 104*, 1004 (1996); O. Quinet and B. Champagne, *Progress in Theoretical Chemistry*, accepted.

[6] Gaussian 94, Revision B.1, M.J. Frisch, G.W. Trucks, H.B. Schlegel, P.M.W. Gill, B.G. Johnson, M.A. Robb, J.R. Cheeseman, T. Keith, G.A. Petersson, J.A. Montgomery, K. Raghavachari, M.A. Al-Laham, V.G. Zakrzewski, J.V. Ortiz, J.B. Foresman, J. Cioslowski, B.B. Stefanov, A. Nanayakkara, M. Challacombe, C.Y. Peng, P.Y. Ayala, W. Chen, M.W. Wong, J.L. Andres, E.S. Replogle, R. Gomperts, R.L. Martin, D.J. Fox, J.S. Binkley, D.J. Defrees, J. Baker, J.P. Stewart, M. Head-Gordon, C. Gonzalez, and J.A. Pople, Gaussian, Inc., Pittsburgh PA, 1995.

[7] T.H. Dunning, *J. Chem. Phys. 90*, 1007 (1989); R.A. Kendall, T.H. Dunning, and R.J. Harisson, *J. Chem. Phys. 96*, 6796 (1992).

[8] G. Maroulis and C. Pouchan, *Phys. Rev. A 57*, 2440 (1998).

TABLE 1: Mean static electronic and vibrational first and second hyperpolarizability contributions of HCN calculated at the RHF and MP2 levels of approximation with the aug-cc-pvtz basis set. All the values are in atomic units (1 a.u. of the first hyperpolarizability = 3.2063 x 10^{-53} C^3 m^3 J^{-2} = 8.641 x 10^{-33} esu; 1 a.u. of second hyperpolarizability = 6.235377 x 10^{-65} $C^4 m^4 J^{-3}$ = 7.0423 x 10^{-54} $m^5 V^{-2}$ = 5.0367 x 10^{-40} esu).
$\beta^K = 3/2\left(\beta^K_{//} - \beta^K_{\perp}\right)$ and $\gamma^K = 3/2\left(\gamma^K_{//} - \gamma^K_{\perp}\right)$ with $\beta^K_{\perp} = \left[2\beta_{\xi\xi z}(-\omega;\omega,0) - \beta_{z\xi\xi}(-\omega;\omega,0)\right]/5$ and $\gamma^K_{\perp} = \left[2\gamma_{\xi\xi\eta\eta}(-\omega;\omega,0,0) - \gamma_{\xi\eta\xi\eta}(-\omega;\omega,0,0)\right]/15$.

	RHF	MP2		RHF	MP2
$\beta^e_{//}(0;0,0)$	-7.3	-0.8	$\gamma^e_{//}(0;0,0,0)$		1799
	-7.6[a]	1.4[a]		188 x 10[a]	217 x 10[a]
$[\mu\alpha]^0_{//;\omega=0}$	1.8	-1.8	$[\alpha^2]^0_{//;\omega=0}$	314	303
$[\mu^3]^{1,0}_{//;\omega=0}$	-4.6	-10.8	$[\mu\beta]^0_{//;\omega=0}$	252	258
$[\mu^3]^{0,1}_{//;\omega=0}$	-5.6	-12.8	$[\mu^2\alpha]^{1,0}_{//;\omega=0}$	527	799
$\beta^v_{//}(0;0,0)$	-8.5	-25.4	$[\mu^2\alpha]^{0,1}_{//;\omega=0}$	-87	-347
$\beta^v_{//}(-\omega;\omega,0)_{\omega\to\infty}$	0.6	-0.6	$\gamma^v_{//}(0;0,0,0)$	1005	1011
$\beta^{v,K}(-\omega;\omega,0)_{\omega\to\infty}$	-5.1	-4.1	$\gamma^v_{//}(-\omega;\omega,0,0)_{\omega\to\infty}$	304	305
			$\gamma^{v,K}(-\omega;\omega,0,0)_{\omega\to\infty}$	240	250
			$\gamma^v_{//}(-2\omega;\omega,\omega,0)_{\omega\to\infty}$	63	65
			$\gamma^v_{//}(-\omega;\omega,-\omega,\omega)_{\omega\to\infty}$	209	202

[a] Ref. 8.

MODE PROPAGATION ALONG THE MULTILAYERED OPTICAL FIBRE WITH KERR LIKE NON–LINEARITY.

E.A. ROMANOVA[1], P. SEWELL[2], T.M. BENSON[2]
1) *Saratov State University, Astrakhanskaya 83, 410026, Saratov, Russia, e-mail: romanova@optics.sgu.ru*
2) *University of Nottingham, Nottingham NG7 2RD, U.K., e-mail: eezps@hermes.nottingham.ac.uk*

Keywords

periodic structure, Kerr–like non–linearity, pulse compression

Abstract

Light propagation in non-linear fibres with multilayered cladding is investigated numerically using the results of modal analysis. The self-focusing of a leaky mode is shown to improve the fibre transmittance resulting in a narrowing of the input field amplitude distribution.

Introduction

Multilayered dielectric two-dimensional structures have been used mainly to lower the losses in hollow metallic waveguides [1]. It has been shown that the addition of dielectric layers could lower the losses significantly. The guiding properties of all–dielectric hollow waveguides were studied recently [2]. Such structures can have lower losses than the combined metal and dielectric structures since the waves do not interact with a lossy metallic layer.
The interaction between a periodic structure and an optical field can be

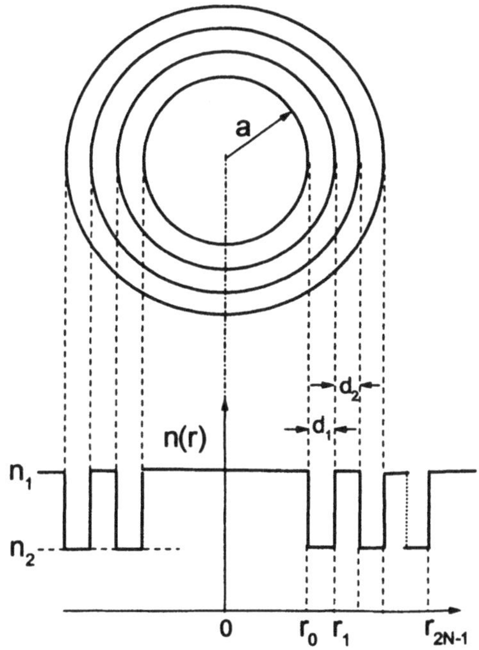

Fig.1

A. Driessen (ed.), Nonlinear Optics for the Information Society, 139–144.

used in novel ways to develop new material systems with desirable properties. One example is the observation of low loss waveguiding in a photonic crystal fibre [3], which only guides a single mode independent of the wavelength and fibre size. This fibre has a core with refractive index equal to that of silica, and a cladding index given by the average index of the periodically holey region which forms the cladding.
In this paper, we consider an all–dielectric fibre having a uniform glass core and a cladding consisting of 2N–1 concentric dielectric layers surrounded by an infinite uniform dielectric cladding having the same refractive index as the core. The cross–section is presented in Fig. 1. Modal analysis of this multilayered structure with linear refractive index has shown that the modal propagation constants are complex in general but tend to become real (i.e. loss-free) as the radiation wavelength is decreased [4].
The continuing development of optical components requires that the behaviour of components including non–linear materials needs to be investigated [5-7]. It was shown in [8] that a Kerr–like non–linearity improved the linear transmittance of a conventional fibre with a step – like discontinuity for some configurations of the structure because a part of the radiation field was self – focused and propagated as a guided mode. This effect can be used to narrow the initial distribution of the field amplitude at the input endface of the fibre.
Here we show the effect takes place when a leaky mode is excited in the multilayered fibre of Fig. 1 with an embedded Kerr–like non–linearity.

Guided and leaky modes of a linear multilayered fibre

To calculate the vector modal field distributions and propagation constants in the linear multilayered fibre, we use the following approach. In each layer, $r_{i-1} < r < r_i$, the longitudinal component of electric field is decomposed [9] into the following set such that:

$$E_z(r) = A_i\ H_n^{(1)}(k_{ti}r) + B_i\ H_n^{(2)}(k_{ti}r) \tag{1}$$

The coefficients A_i, B_i are derived in [7], $H_n^{(i)}(x)$ is the Hankel function of the n th order of the i th kind and $k_{ti}^2 = n_i^2 k_o^2 - \beta^2$, where β is a longitudinal propagation constant. A similar expansion can be found for the longitudinal component of magnetic field $H_z(r)$. It is noted that in the core (r < a) and outer cladding region (r > a $+(d_1 + d_2)N - d_2$) the expression (1) simplifies to:

$$E_z(r) = \frac{E_z(r_1)}{J_n(k_{t1}r_1)} J_n(k_{t1}r)\ , E_z(r) = \frac{E_z(r_N)}{H_n^{(2)}(k_{tN}r_N)} H_n^{(2)}(k_{tN}r), \tag{2}$$

respectively.
From these expressions it is straightforward to derive the tangential field components E_θ and H_θ on each side of the boundaries between the layers in terms of the values of E_z and H_z at the boundaries. Requiring continuity of E_θ and H_θ yields a matrix equation whose determinant will only be zero if β is a modal propagation constant. Two different linear structures were considered using this method. These had low and high index

contrast in the cladding (Fig.2) respectively with a = 2.5μm, d_1 = 0.5μm, d_2 = 0.5μm, N = 8.

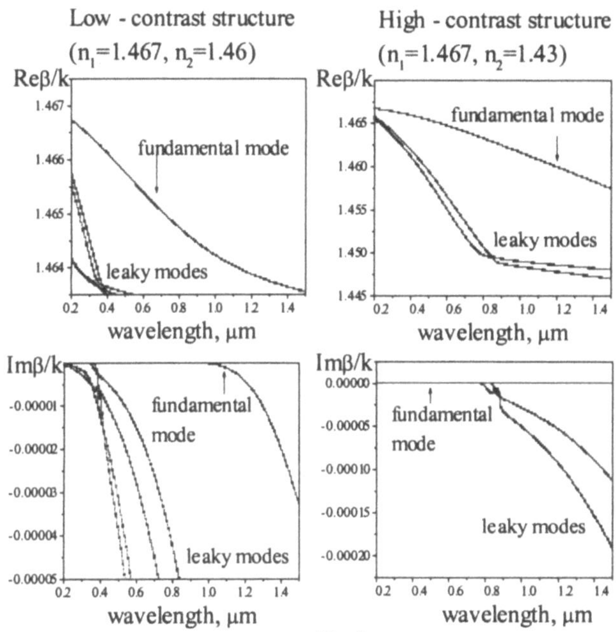

Fig.2

Two types of complex solutions were identified: the modes with Imβ/k < 0.1dB km^{-1} which can be treated as the guided ones and a variety of closely located roots corresponding to the leaky waves. The fundamental mode was shown to behave as the guided one over a wide range of wavelengths depending on the refractive index contrast (n_1 - n_2). The leaky modes transformed into the guided ones as the wavelength decreased.

Mode transformation in a non – linear multilayered fibre

We have analysed the self-focusing of the field propagating along the fibre axis using the Finite-Difference Beam Propagation Method (FD-BPM) [10], for the low – contrast structure excited numerically by a Gaussian beam at a wavelength λ and with the beam halfwidth equal to the fibre radius. The Kerr constant n_2 was assumed invariant over the fibre cross – section.

The scalar wave equation for the slowly varying amplitude of the transverse component of the electric field *F(r,z)* was transformed into a finite-difference scheme using a

Crank-Nicholson three – point approximation and a recursive procedure known as a sweep method [11]. A split – step approximation was used in such a way that the non – linear susceptibility was calculated first for the given field profile $\chi_i = \chi^{(0)} + 2a^2k^2n(r)n_2|F_{i-1}|^2$ with $k = 2\pi/\lambda$, Kerr – constant n_2 and linear susceptibility $\chi^{(0)}$. This distribution was then used in the finite – difference scheme to evaluate the field profile F_i. The longitudinal power flow $S_z(z)$ was calculated by averaging the longitudinal component of the Poynting vector over the structure cross – section. The boundary conditions used were: $F(\infty,z) = 0$, $\partial F(0,z)/\partial r = 0$. To avoid non–physical reflections from the workspace boundaries, complex scaling was applied [12]. The numerical errors were no more than $10^{-5}\%$ for the following parameters of the computational window:

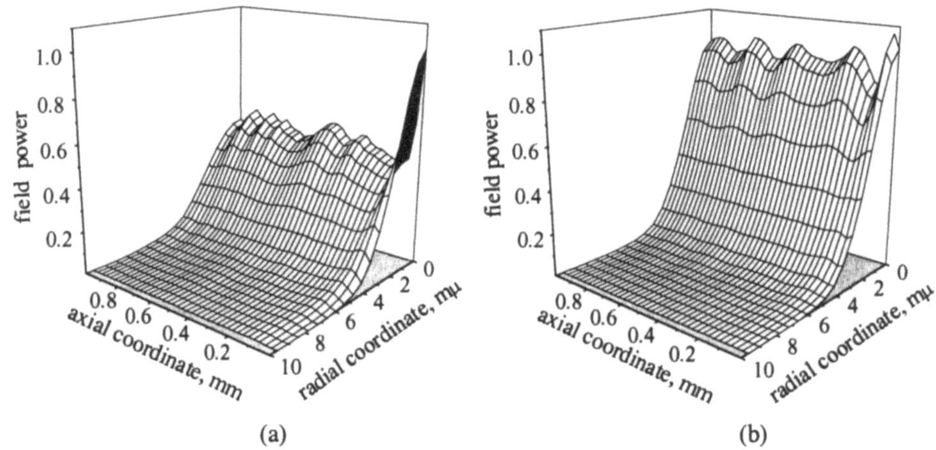

Fig.3

mesh size $\Delta r = 0.005\mu m$, longitudinal step size $\Delta z = 7.0\mu m$. The number of transverse mesh points was 10^4.

In the presence of a Kerr–like non–lineaity ($n_2S_z(z) \approx 10^{-4}$) the guiding properties of the structure increase due to the self – focusing. A leaky mode ($\lambda = 1\mu m$) of the linear structure (Fig.3, a) has a tendency to become a guided one in the nonlinear structure (Fig.3, b).

The transmittance $\tau_{nl}(L) = S_z(L)/S_z(0)$ of the non–linear multilayered fibre depends on the input field amplitude. It can be greater or less than the linear transmittance $\tau_l(L)$ depending on the beam wavelength (Fig.4). The nonlinear transmittance of a leaky mode ($\lambda = 1.0\mu m$, $\lambda = 1.3\mu m$, Fig.2) is greater than the corresponding linear one. The effect is stronger at larger wavelengths, which correspond to leaky modes with greater attenuation coefficient. Otherwise, a guided mode ($\lambda = 0.7\mu m$, Fig.2) losses its power when it transforms into the "non – linear" one. This result can be considered as an example of the mutual action of the field leakage and non–linearity of the fibre material resulting in a better transmittance for the more powerful light beam. An analogous

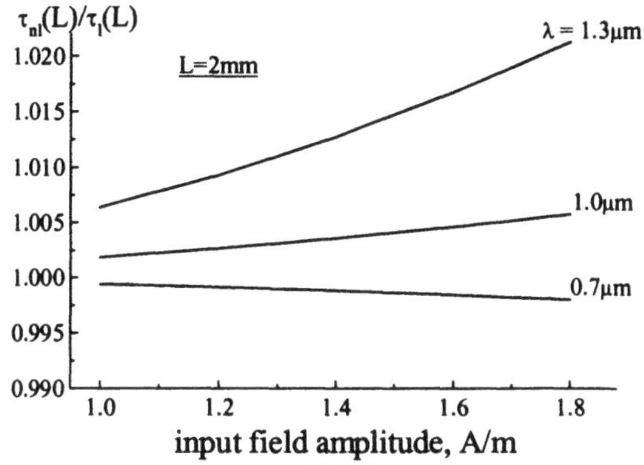

Fig.4

effect has been considered in [8] where an irregular waveguide consisting of two segments of conventional step–index fibre having different core radii was modelled.

Conclusion

Non–linear all–dielectric fibres with a uniform core and multilayered cladding have been investigated as potential optical elements useful to control the propagation of light. In particular, the self – focusing effect is shown to improve the longitudinal transmittance of such structures, resulting in a narrowing of the input field amplitude distribution. The silica–fibre non–linearity is known to be relatively small ($\approx 10^{-16}$ cm^2/W). However, novel materials with larger Kerr coefficients and response times of the order of picoseconds are available now and the peak pulse intensities can be as high as some GW/cm^2 [13]. This is why the non–linear effect, considered herein, can be used to devise pulse compressors and active elements of fibre lasers.

References

[1] M.Miyagi and S.Kawakami, Design theory of dielectric coated circular metallic waveguides for infrared transmission, *J. Lightwave Technol. LT-2*, pp.116-126, 1984.

[2] Y.Fink, D.J.Ripin, S.Fan, C.Chen, J.D.Joannopoulos and E.L.Thomas, Guiding optical light in air using an all-dielectric structure, *J. Lightwave Technol. LT-17*, pp. 2039-2041, 1999.

[3] J.C.Knight, T.A.Birks, R.F.Cregan, P.St.J.Russell and J.- P. de Sandro, Photonic crystals as optical fibres – physics and applications, *Optical materials 11*, pp.143 – 151, 1999.

[4] E.A.Romanova, P.Sewell and T.M.Benson, Wave propagation through all-dielectric multilayered optical fiber, *Proceedings of the 2nd International Conference on Transparent Optical Networks*, M.Jaworski (ed.), National Institute of Telecommunications, Warsaw, *IEEE Catalog Number 00EX408, ISBN 0-7803-6337-X*, pp. 57-60, 2000.

[5] N.D.Sankey, D.F.Prelewitz and T.G.Brown, All – optical switching in a nonlinear periodic waveguide structure, *Appl.Phys.Lett. 60*, pp.1427 – 1429, 1992.

[6] S.Radic, N.George and G.P.Agrawal, Theory of low – threshold optical switching in non – linear phase – shifted periodic structures, *J.Opt.Soc.Am. B, 12*, pp.671 – 680, 1995.

[7] H-B.Lin, R.J.Tonucci, and A.J.Campillo, Two–dimensional photonic bandgap optical limiter in the visible, *Opt.Lett.23*, pp.94 – 96, 1998.
[8] E.A. Romanova, L.A. Melnikov and E.V. Bekker, Scattering of the total field from the slow – tapered and step – like discontinuities of dielectric waveguides, *Microwave Opt. Technol. Lett. 25*, pp. 27 – 33, 2000.
[9] R.E.Collin, Field Theory of Guided Waves. *IEEE Press.1991*.
[10] W.P.Huang (ed.), Methods for modelling and simulation of guided wave optoelectronic devices: waves and interactions, *PIERS 11*, EMW Publishing, Cambridge, Massachusetts, USA, 1995.
[11] A.N.Tichonov and A.A.Samarsky, Equations of Mathematical Physics, *"Nauka"*. Moskva.1972 (in Russian).
[12] C.W.McCurdy and C.K.Stroud, "Eliminating wavepacket reflection from grid boundaries using complex coordinate contours", *Computer Physics Communication 63*, pp. 323-330, 1991.
[13] C.M. de Sterke, B.J.Eggleton,P.A.Krug, "High-intensity pulse propagation in uniform gratings and grating superstructures", *J. Lightwave Technol. LT-15*, pp.1494-1502, 1997.

NUMERICAL STUDY OF OPTICAL BISTABILITY IN A NONLINEAR PHOTONIC CRYSTAL NOTCH FILTER

R. STOFFER[1], Y.S. KIVSHAR[2]

1) *Mesa+ Research Institute, University of Twente, P.O. Box 712, 7500 AE Enschede, The Netherlands Phone: +31-53-4892743; fax : +31-53-4893343; e-mail: r.stoffer@el.utwente.nl*
2) *Optical Sciences Centre, RSPhysSE, Australian National University, Canberra, ACT, 0200, Australia Phone: +61-2-62492430; fax: +61-2-62495184; e-mail: ysk124@rsphy1.anu.edu.au*

Abstract

Optical bistability occurs when the effects of nonlinear behaviour of materials cause hysteresis in the transmission and reflection of a device. A possible mechanism for this is a strong dependence of the optical intensity on the index of refraction, e.g. in a cavity near resonance. In a 2-dimensional photonic crystal composed of rods of high-index material in air, a waveguide can be created by removing a line of rods. When a cavity is made by taking away several rods perpendicular to the waveguide, a notch filter characteristic in the transmission occurs. Due to the high intensity in the cavity in resonance, nonlinear effects are enhanced. This paper shows numerical simulations of bistability in the transmission and in the field inside the cavity both when a material inside the cavity has third-order (Kerr-type) nonlinear effects, and when the high-index rods themselves are nonlinear.

Full article not available

A. Driessen (ed.), Nonlinear Optics for the Information Society, 145.

A LOW-DIMENSIONAL MODEL FOR DEFORMATION OF BICHROMATIC WAVES IN THIRD ORDER NONLINEAR MEDIA

A. SURYANTO, E. VAN GROESEN, H.J.W.M. HOEKSTRA
University of Twente
P.O. Box 217, 7500 AE, Enschede, the Netherlands

Abstract

It is well known that the propagation of optical pulses and beams in material with third order nonlinearity can be modelled by the NLS equation. In this paper we present a low-dimensional model for the deformation of bichromatic pulses (beams). Motivated by the cubic nonlinearity in the NLS equation, the model is constructed in a 4D manifold of the complex coefficients of the two most dominan harmonic temporal modulation. By exploiting the conservation of energy and gauge, the model is reduced to a 2D Hamiltonian system. Numerical solutions of the system show that the present model describes the correct qualitative deformations. Furthermore, as long as higher frequency temporal modulations are ngeligible, the description is also quantitavely correct.

Full article not available

A. Driessen (ed.), Nonlinear Optics for the Information Society, 147.

PICOSECOND ABSORPTION SATURATION DYNAMICS OF NEW $[M(R,R'timdt)_2]$ METAL-DITHIOLENES

M.C. ARAGONI[1], M. ARCA[1], T. CASSANO[2], F.A. DEVILLANOVA[1], M. FERRARA[2], R. TOMMASI[2,3]

1) *Dip. di Chimica Inorganica ed Analitica, Università di Cagliari, S.S. 554 bivio Sestu, I-09042 Monserrato-Cagliari, Italy*
2) *INFM and Dip. Di Fisica, Università di Bari, Via G. Amendola 173, I-70126 Bari, Italy*
3) *DETO, Università di Bari, Via G. Amendola 173, I-70126 Bari, Italy*

Keywords

Dithiolenes, Z-scan, nonlinear absorption, picosecond spectroscopy

Abstract

The nonlinear absorption properties of new fluorinated neutral metal-dithiolenes belonging to the general class $[M(R,R'timdt)_2]$ (M = Ni, Pd, Pt; R,R'timdt = monoreduced form of disubstituted imidazolidine-2,4,5-trithione) have been studied. The absorption has been found to saturate with the incident light intensity and the temporal evolution of the absorption saturation has been investigated by means of the picosecond time-resolved pump-probe technique. We observed a strong dependence of both the saturation intensity and the temporal dynamics on the metal ion.

1. Introduction

Among nonlinear optical materials with possible applications in the field of laser technology [1] metal-dithiolenes represent a very promising class of compounds, because of their intense absorption in the Visible-NIR region, and of their thermal and photochemical stability [2].
Recently, some of us have reported the new class of neutral metal dithiolenes $[M(R,R'timdt)_2]$ (M = Ni, Pd, Pt; R,R'timdt = monoreduced form of disubstituted imidazolidine-2,4,5-trithione [3][4] having a very high degree of π-electron delocalization extended over the whole molecule, which results in an unusually intense NIR absorption (ε as high as 90000 M^{-1} cm^{-1}). Moreover, the changes in the metal ion and/or the R and R' substituents allow to tune the position of the absorption maximum λ_{max} between 990 and 1030 nm [4], the absorption at the highest wavelength being

A. Driessen (ed.), Nonlinear Optics for the Information Society, 149–154.

obtained when R' = aromatic substituents and M = Pd. The strength and the spectral position of their absorption maxima candidate this class of dithiolenes as dyes for Q-switching and/or mode-locking Neodymium lasers (wavelength in the range 1053-1064 nm) [5]. Nonetheless, the solubility of these complexes is still quite low to make them useful for real applications.

2. Results and discussions

In order to increase the solubility of metal dithiolenes in organic solvents and to bring their λ_{max} values as close as possible to the Nd laser emission wavelength, we synthesised the systems 1-3 [R = Et; R' = 4-trifluoromethoxyphenyl; M = Ni (1), Pd (2) and Pt (3)], which bear a polar CF_3 group bound to the aromatic ring (Scheme 1).

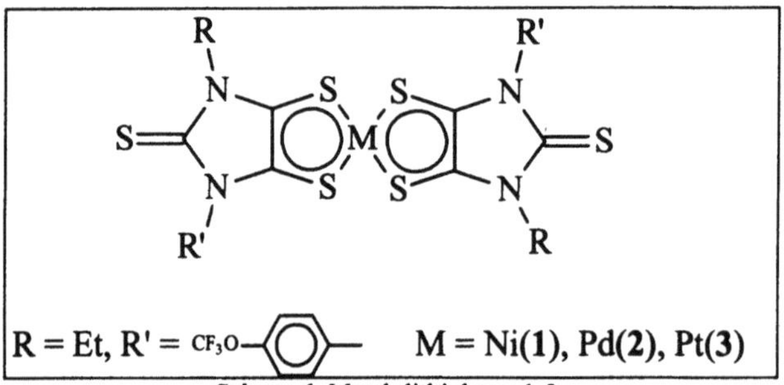

Scheme 1: Metal-dithiolenes 1-3.

According to what was previously reported [3][4], syntheses were accomplished by reacting Lawesson's reagent [6] and *N*-ethyl-*N'*-(4-trifluoromethoxyphenyl)-imidazolidine-2-thione-4,5-dione in the presence of the metal, introduced as metal powder (M = Ni) or as chloride (M = Pd, Pt). The λ_{max} position follows the trend already observed [4] on passing from Ni to Pt [λ_{max} = 1010 nm (1), 1032 nm (2), 1010 nm (3)] and all complexes are characterised by a very intense NIR absorption. In particular, complex 3 exhibits the highest extinction coefficient (ε = 91000 M^{-1} cm^{-1}) ever reported for a metal-dithiolene (see Fig. 1).

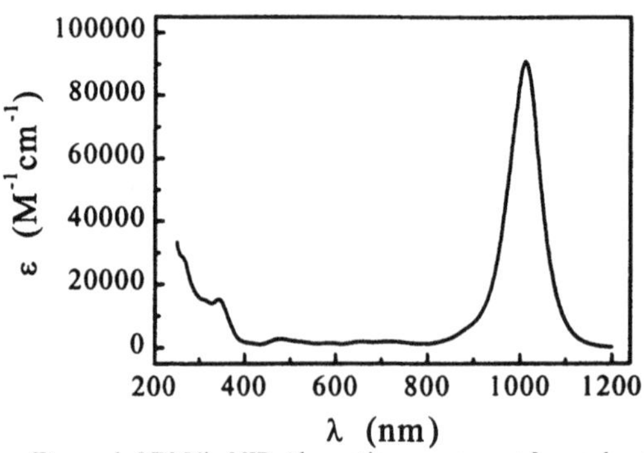

Figure 1: UV-Vis-NIR Absorption spectrum of complex 3.

To investigate the optical nonlinearities of these systems in the NIR region, we performed absorption measurements at different incident light intensities. The dithiolenes were dissolved at a concentration $c \cong 4.2 \cdot 10^{-5}$ M in chloroform and placed in a 1-mm-thick quartz cuvette. The solutions were excited at room temperature using a mode-locked and Q-switched Nd:YAG laser delivering $\lambda = 1064$ nm , $\tau = 35$ ps (FWHM) pulses at a 10 Hz repetition rate. The maximum pulse energy ($E_{max} \cong 20$ mJ) was appropriately reduced using calibrated neutral density filters. The sample transmittance as a function of the incident laser intensity was recorded using a double-beam setup. The laser beam was split into two parts: one was reflected and sent through a cuvette containing only the solvent and used as a reference, the other transmitted through the sample. The signals were collected using a data acquisition system and each experimental point is the average of 100 measurements. All the experiments were performed at room temperature.
The absorption coefficient as a function of the incident laser intensity for sample 1 is plotted in Fig. 2. This clearly shows that the dithiolene behaves like a saturable absorber and that its absorption coefficient can be written as:

$$\alpha = \frac{\alpha_0}{1 + I/I_s}, \quad (1)$$

where α_0 is the linear absorption coefficient, I the excitation intensity and I_s the saturation intensity, i.e. the intensity at which the absorption coefficient is reduced to one half of its unsaturated value α_0.
Similar results have been obtained for dithiolenes 2 and 3. A simple fit of the experimental data with Eq.(1) gives I_s~450 MW/cm^2, ~250 MW/cm^2, and ~350 MW/cm^2 for complexes 1, 2 and 3, respectively.

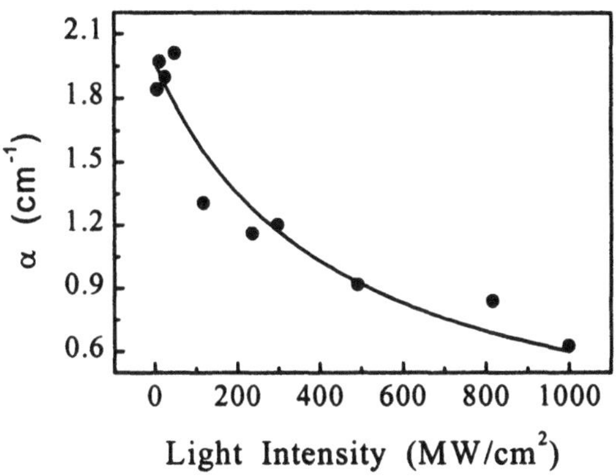

Figure 2: Absorption coefficient (dots) of complex 1 as a function of the incident laser intensity. The solid line is the fit obtained using Eq. 1.

To get a deeper insight into the absorption nonlinearities of our systems, we investigated their absorption saturation dynamics on a picosecond time scale. Measurements were performed using the pump-probe technique where dithiolenes are excited to an upper

electronic state by a pump pulse and their subsequent evolution is followed by measuring the induced transmission changes of a time-delayed probe pulse.
For an optically thin sample, the absorption change $\Delta\alpha$ is proportional to the differential transmission $\Delta T(t)/T$:

$$\frac{\Delta T(t)}{T} \propto -\Delta\alpha(t) \times L, \tag{2}$$

where T is the probe transmission, ΔT is its variation due to the pump pulse, t is the time delay between the pump and probe pulses and L the sample thickness. The measured normalised $\Delta T(t)/T$ at $\lambda = 1064$ nm are reported in Fig. 3 for samples 1 and 3.
For all the investigated dithiolenes, after a time-resolution limited increase, the transmission of the complexes decreases bi-exponentially. To obtain more quantitative information on the absorption saturation dynamics, we have fitted our data using a simple phenomenological response function R:

$$R(t) = u(t)\{I - D_1[1 - \exp(-t/\tau_{d1})] - D_2[1 - \exp(-t/\tau_{d2})]\}, \tag{3}$$

where u(t) is the unit step function. The first term accounts the assumed instantaneous increase of the sample transmission due to $S_0 \rightarrow S_1$ absorption saturation. The second and the third terms describe the relaxation of the system. Convolution of this response function with the pump-probe cross correlation has been fitted to our data (solid lines in Fig. 3).

The fast decay time is constant (within the precision of the experiment) for the three dithiolenes and amounts to $\tau_{d1} \sim 15$ ps. In contrast, the slow decay time τ_{d2} strongly depends on the central metal ion. In particular, on passing from Ni to Pt a dramatic shortening of the dye recovery time is observed. The least-squares fit of our data gives $\tau_{d2} \sim 12000$ ps (1), ~280 ps (2), and ~200 ps (3).

3. Conclusions

Our results clearly indicate that both the saturation intensity and the bleaching temporal dynamics of our systems are largely influenced by the metal ion and can be tuned in wide ranges. This, together with their high thermal and photo-chemical stability makes dithiolene complexes belonging to the $[M(R,R'timdt)_2]$ class promising candidates for stabilising soliton-like pulses [7] or to achieve efficient passive mode-locking [8] at wavelengths around one micron.

Acknowledgements

We wish to thank Romina Frau for help in the synthesis, “Progetto Finalizzato Materiali Speciali per Tecnologie Avanzate II” of the “Consiglio Nazionale delle Ricerche” and MURST (“Ministero dell’Università e della Ricerca Scientifica e Tecnologica”) for funding.

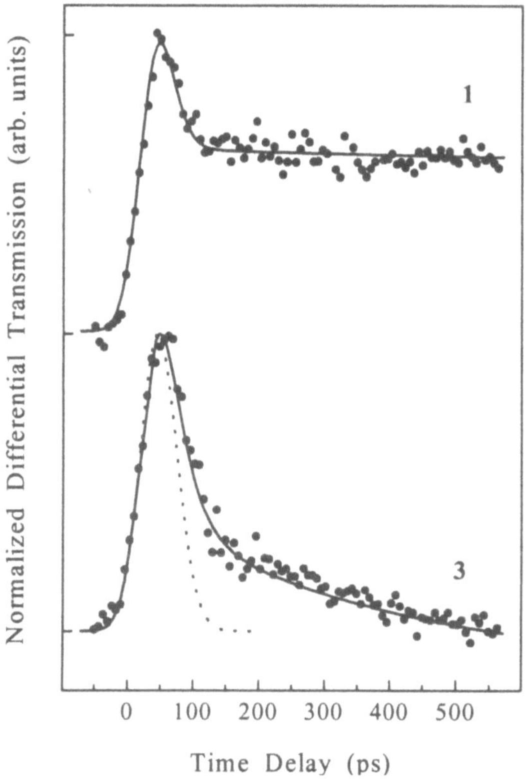

Figure 3: Experimental ΔT/T (dots) as a function of the pump-probe time delay in samples 1, 3 for identical pump and probe wavelengths (1064 nm). The solid lines are the least-squares fits of the experimental data using Eq. 3. The dashed line is the cross-correlation of the pump and probe laser pulses.

References

[1] *Molecular Nonlinear Optics: Materials, Physical and Devices, ed. J. Zyss*, Academic Press, Boston, MA, 1993; *Materials for Nonlinear Optics – Chemical Perspectives*, ed. S. R. Marder, J. E. Sohn ad G. D. Stucky, ACS Symp. Ser., Americal Chemical Society, Washington DC, 1991; *Introduction to Nonlinear Optical Effects in Molecules and Polymers*, ed. P. N. Prasad and D. J. Williams, Wiley, New York, 1991; *Nonlinear Optical Materials*, ed. H. Kuhn and J. Rubillard, CRC Press, Boca Raton, FL, 1992.

[2] U.T. Mueller-Westerhoff, B. Vance and D.I. Yoon, *Tetrahedron* 1991, 47, 909; J. A. McCleverty, *Prog. Inorg. Chem.*, 1968, 10, 49

[3] F. Bigoli, P. Deplano, F.A. Devillanova, V. Lippolis, P.J. Lukes, M.L. Mercuri, M.A. Pellinghelli and E.F. Trogu, *J. Chem. Soc., Chem. Commun.* 1995, 371; F. Bigoli, P. Deplano, F. A. Devillanova, J. R. Ferraro, V. Lippolis, P. J. Lukes, M. L. Mercuri, M. A. Pellinghelli and E. F. Trogu, *Inorg. Chem.* 1997, 36, 1218; M. Arca, F. Demartin, F. A. Devillanova, A. Garau, F. Isaia, F. Lelj, V. Lippolis, S. Pedraglio an G. Verani, *J. Chem. Soc. Dalton Trans* 1998, 3731.

[4] M.C. Aragoni, M. Arca, F. Demartin, F.A. Devillanova, A. Garau, F. Isaia, F. Lelj, V. Lippolis and G. Verani, *J. Am. Chem. Soc.*, 1999, 121, 7098.

[5] K.H. Drexage and U.T. Mueller-Westerhoff, *IEEE J. Quantum Electron.*, 1972, QE-8, 759

[6] S. Sheibe, B.J. Pedersen and S.O. Lawesson, *Bull. Soc. Chim. Belg.*, 1978, 87, 229.

[7] F.X. Kärtener and U. Keller, *Opt. Lett.*, 1995, 20, 16.
[8] J.L.A. Dubbeldam and D. Lenstra, *Appl. Phys. Lett.*, 1999, 75, 1066.

MEASUREMENTS OF THE NONLINEAR COEFFICIENT N_2/A_{EFF} USING A SELF-ALIGNED INTERFEROMETER AND A FARADAY MIRROR

C. VINEGONI, M. WEGMULLER, N. GISIN
Group of Applied Physics, University of Geneva, 20 Ecole-de-Medecine, CH-1211 Geneve 4, Switzerland

Abstract

A simple, all-fiber implementable method for the measurement of the nonlinear coefficient n_2/A_{eff} in telecom fibers at 1550 nm is demonstrated. The method is based on the interferometric detection of the Kerr phase shift acquired by a laser pulse along the fiber under test. The detection is made by an all-fiber self-aligned interferometer incorporating a Faraday mirror. The self alignment characteristic allows for an easy and quick initial adjustment of the interferometer and leads to a good robustness as the two interferometer arms are always automatically matched. The Faraday mirror, whose property is that it transforms any input polarization state to the orthogonal one upon reflection, completely removes any polarization transformation of the fiber under test. In the same way, any fluctuations of the polarization state due to environmental perturbations (temperature, pressure changes) are removed. The polarization state re-entering the self-aligned interferometer on the backwards path is therefore fixed, so that it can be adjusted once-for-all for maximum visibility at its output. The presence of the Faraday mirror makes the setup robust, leading to a good accuracy in the n_2/A_{eff} value. Moreover, the fiber under test can be easily replaced without necessitating any further readjustment of the interferometer. The proposed scheme is therefore well suited to routinely measure the nonlinear coefficient.

1. Introduction

The implementation of Erbium-doped fiber amplifiers and chromatic dispersion compensation allows for long distance data transmission. Along with the technique of wavelength division multiplexing (WDM), this leads to an important amount of power inside the fiber over long distances, and optical nonlinearities start to play a significant role. Their magnitudes depend on the ratio n_2/A_{eff}, where n_2 is the nonlinear refractive index of the fiber and A_{eff} the effective area of the lightmode. There are different methods to measure n_2/A_{eff}, based on SPM or XPM [1] induced phase shift detection using interferometric and non-interferometric schemes. The interferometric detection scheme [2] has the advantage that it can be implemented more easily - a disadvantage is however its susceptibility to environmental perturbations leading to a poor stability. A

A. Driessen (ed.), Nonlinear Optics for the Information Society, 155–159.

considerable improvement of this technique, is presented in the following, using a self-aligned interferometer [3] with a Faraday mirror. The suggested method has the advantage to be simple and all fiber implementable, and the fluctuations due to environmental perturbations are completely removed.

2. Principle of operation

Due to the power dependence of the refractive index (optical Kerr effect), the phase acquired by a pulse with peak power P and wavelength λ that travels through a fiber of length L is given by

$$\phi(P) = \phi_l + \phi_{nl} = n_0 kL + kL_{eff}\frac{n_2}{A_{eff}}Pm \qquad (1)$$

The decrease in pulse power from the fiber attenuation is accounted for by L_{eff}. The polarization parameter m depends on the polarization characteristics of the test fiber and the input signal polarization state. For the case of a high birefringent fiber and light coupled into one axis only, m = 1. For the case of a long standard telecom fiber with a complete polarization scrambling, it was demonstrated that m = 8/9 [4]. Measuring the acquired phase shift, Eq.1 allows to determine the ratio n_2/A_{eff}.
The setup of the self-aligned interferometer we have used in order to measure this phase shift is shown in Fig.1. High peak power pulses from an EDFA are split at the entry of the first coupler (coupling ratio 50/50) and move along the two arms of the interferometer. These arms are different in length such that the two pulses do not interfere when they recombine at the second coupler (coupling ratio 90/10). Due to the asymmetry of the coupler, the two pulses enter the fiber under test (FUT) with different powers, and according to Eq.1, they will acquire different amounts of phase shift during propagation. The pulses are then reflected at the Faraday mirror (FM) and return back through the FUT and interferometer towards the first coupler. Four different paths are possible during the go- and return trip through the interferometer. A double pass of the long-arm (LL), of the short-arm (SS), and a forward pass of the short (long) arm with a return pass through the opposite arm (SL and LS, respectively). Because of the different arm lengths, three different arrival times at the detector can be discerned. Only the middle pulse, arising from the interference between the SL and the LS pulses, is analyzed. The power of this pulse depends on the phase relationship between the two interfering signals and it can be exploited to calculate the nonlinear phase shift experienced in the FUT. It is important to note that contrary to regular Mach-Zehnder interferometers, the balancing of the interferometer arms is not critical here as the path lengths of the two interfering signals are automatically matched (self-aligned).

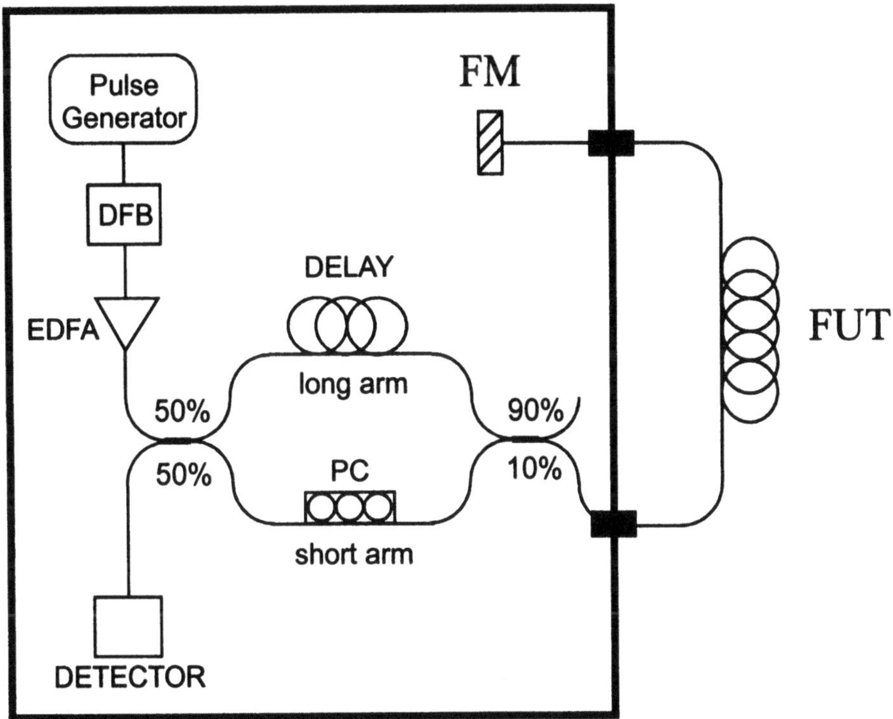

Figure 1. Experimental setup of the self-aligned interferometer. DFB distributed feedback laser, EDFA Erbium doped fiber amplifier, PC polarization controller, FUT fiber under test, FM Faraday mirror, D detector

To calculate the detected power of the middle pulse we have to consider the state of polarization (SOP) of the two interfering pulses. Let R_L, R_S, and R_{FUT} be the polarization transformation operators for the long and short arm of the interferometer and for the FUT, respectively. Knowing that the action of the FM is such that upon reflection, the SOP of the reflected light is always orthogonal to the incoming one, we can describe the double pass of the FUT as

$$m_g = R_{FUT}^{-1} - FR_{FUT}m_0 = R_{FUT}^{-1}(-(R_{FUT}m_0)) = -R_{FUT}^{-1}R_{FUT}m_0 = -m_0$$

So for the pulses that travel through the LS (SL) path, the SOP m_{LS} (m_{SL}) at the exit of the interferometer is given by:

$$m_{LS} = R_S^{-1}\left(R_{FUT}^{-1}FR_{FUT}\right)R_L m_0 = R_S^{-1}(-R_L m_0) = -R_S^{-1}R_L m_0 = Am_0$$

$$m_{SL} = R_L^{-1}\left(R_{FUT}^{-1}FR_{FUT}\right)R_S m_0 = R_L^{-1}(-R_S m_0) = -R_L^{-1}R_{SL} m_0 = Bm_0$$

To have maximum visibility at the exit of the interferometer the two pulses need to have the same SOP, i.e. A=B. This can be done by adjusting the polarization controller (PC) inserted in one of the interferometer arms, e.g. such that $R_S=R_L$. Note that using a standard mirror in place of the FM, an additional PC would be required [5], making the initial adjustment of the interferometer more difficult. Moreover, both PCs would have to be readjusted for every new FUT. With the FM instead, the interferometer does not require any adjustments after its initial calibration.

Another important point is the possibility to adjust the bias of the interferometer. In the experiment, the PC was adjusted in such a way that the above requirement of A=B was fulfilled. Consequently, all the light was directed towards the detector at the first coupler.

Taking into account this and Eq. (1), the detected power P_{OUT} for a standard fiber becomes

$$P_{OUT}(P) \propto P\cos^2(\Delta\phi) \tag{2}$$

where $\Delta\phi$ is the acquired nonlinear phase shift [5],

$$\Delta\phi(P) = \frac{2\pi}{\lambda} PL_{eff} \frac{16}{45} \frac{n_2}{A_{eff}} \tag{3}$$

3. Results

The pulse source used in the experiment was a directly modulated DFB laser diode. Its wavelength was 1559 nm, the pulse duration 28 ns, and the repetition rate 1 kHz. Note that in some fibers, such long pulses can excite acoustic waves through electrostriction, leading to an erroneous measurement of n_2/A_{eff}.

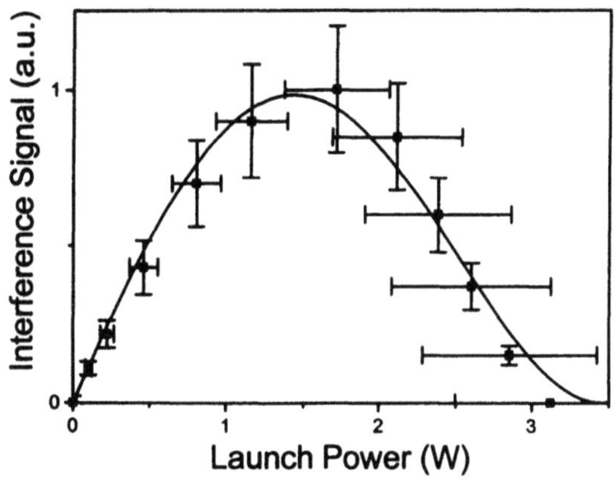

Figure 2. Detected interference signal power as a function of launch power: (squares) measured data, (solid line) theoretical fit

The results for a standard telecom fiber with a length of 1100 m are shown in Fig.2. The interference signal power is plotted as a function of the launch power P. The experimental values (squares) are increasing almost linearly in the beginning, demonstrating that nonlinear effects are of little importance up to about 0.8 W of launched power. Then they set in quite heavily, and the measured power eventually starts to decrease with increasing launch power. The maximum of the interference signal power is reached at a lunch power of 1.4 W, whereas a null value, corresponding to a full $\pi/2$ nonlinear phase shift, is obtained for 3.1 W. From this value, n_2/A_{eff} can be calculated using Eq.2. The precision is however much better if the whole curve is fitted. This fit is represented by the solid line in Fig.2. As can be seen, the measured data and the fitted values correspond well, with $\chi^2=3x10^{-3}$. From the fit we obtain a value of $(2.76\pm0.04)\ 10^{-10}\ W^{-1}$ for the nonlinear coefficient n_2/A_{eff}.

4. Conclusion

We have presented a simple method for the measurement of the nonlinear coefficient n_2/A_{eff} based on an all fiber, self-aligned interferometer. Self-alignment not only allows for an easy and quick initial adjustment of the interferometer, but along with the use of a Faraday mirror also makes it robust against environmental perturbations. This leads to a good accuracy for the measured n_2/A_{eff} values. The proposed method is well suited to routinely measure the nonlinear coefficient, as the fiber under test can be easily exchanged without necessitating any further readjustments of the interferometer.

Acknowledgements

We acknowledge the financial support from the Swiss Federal Office for Education and Sciences (OFES) in the framework of the European COST 265 action.

References

[1] A. Melloni, L. Amoto, P. Boffi, A. Fellegara, P. Sacchetto, M. Martinelli, Characterization of optical Kerr coefficient in telecommunication fibers, by a polarization insensitive method, *EQEC 1996* p. 102

[2] F. Wittl, J. Vobian, G. Herchenroeder, W. Dultz, Interferometric determination of the nonlinear refractive index n2 of optical fibers, *Technical Digest - Symposium on Optical Fiber Measurements, (NIST SP 905)*, 1996, pp. 71–74

[3] J. Meier, Stabile Interferometrie des nichtlinearen BrechzahlKoeffizienten von Quarzglasfasern der optischen Nachrichtentechnik, *Ph.D. Thesis* (1995)

[4] S.V. Chernikov, J.R. Taylor, Measurement of normalization factor of n_2 for random polarization in optical fibers, *Opt. Lett. 1996, 21*, pp. 1559–1561

[5] C. Vinegoni, M. Wegmuller, N. Gisin, Determination of the nonlinear coefficient n_2/A_{eff} using a self-aligned interferometer and a Faraday mirror, *Elec. Lett.,Accepted for publication.*

NONLINEAR OPTICS FOR QUANTUM COMMUNICATION

H. ZBINDEN
Université de Genève, GAP-O 20
Rue de l'Ecole-de-Médecine, CH-1211 Genève 4

Abstract

Presently, Quantum Communication is a very hot topic. People are dreaming about quantum computers and teleportation. However, the experimental problems are huge if not insurmountable and quantum cryptography is the only practical application that emerges so far.
We give a brief introduction in quantum cryptography, quantum computing and teleportation. In many experiments nonlinear optics is involved in a way or another. We discuss what kind of developments in nonlinear optics would be fruitful for experiments related to Quantum Communication.

Full article not available

A. Driessen (ed.), Nonlinear Optics for the Information Society, 161.

MULTISTABILITY AND HYSTERESIS OF STANDING WAVES IN A PERIODIC STRUCTURE

W.A.C. STRIJBOSCH, T.P. VALKERING, S.A. VAN GILS
Mesa[+] Research Institute, Faculty of applied physics, University of Twente,
P.O. Box 217, 7500 AE Enschede, The Netherlands

Abstract

An experimental set-up is presented to create standing nonlinear continuous waves in a Kerr wave guide with a periodic structure. The frequency of the wave is in the band gap of the system. Control parameters are the intensity of the source at one side of the structure, and the phase difference between the forward and backward moving components of the standing wave at the other side. For a simplified system, a bilayer grating, it is shown that multi-valuedness of the solutions occurs with respect to these two control parameters. To obtain these results, analytical and numerical methods are used to describe the spatial 'dynamics' of the nonlinear transfer map of one unit of the grating. Among other things, it is demonstrated that effects of chaotic dynamics can be visible in realistic systems.

1. Introduction

Nonlinearity is known to cause a variety of interesting effects in many optical systems. In gratings one finds envelope solitons, multi-valuedness and transparency for frequencies within the band gap. Most attention and experiments so far are devoted to *propagating* waves, with a nonzero energy flux through the system. This paper is devoted to nonlinear *standing* continuous waves with frequency in or near the band gap. In such standing waves the average energy flux is zero.

AN EXPERIMENTAL SETUP

Consider a situation as shown in Figure 1. A source sends a continuous wave with amplitude E and frequency ω through a waveguide with a periodic structure. The waveguide is made from Kerr material. The waves are reflected completely at a mirror, so that the resulting excitation in the system has zero energy flux. This results in a standing wave. Since the electric field is zero at the (metallic) mirror, the optical path length between mirror and grating determines partially the boundary condition of the electric field at the mirror side of the grating. This optical path length will be one of the

A. Driessen (ed.), Nonlinear Optics for the Information Society, 163–178.

control parameters. The other control parameters are the intensity of the source and, obviously, its frequency.

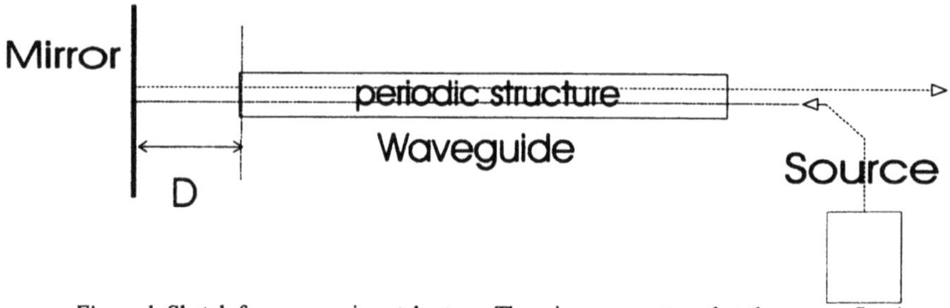

Figure 1: Sketch for an experimental set-up. The mirror guarantees that the energy flux is zero.

For an experimental situation as sketched we consider in particular the multi-valuedness of the solutions. Multi-valuedness is observed usually when a finite grating is illuminated from one side. Then for one value of the incoming intensity, more than one value of the transmitted intensity is possible. In the present case multi-valuedness appears in a different way: for a range of values of intensities of the source and of the optical path length between mirror and grating, one finds that the field in the grating has more than one possible realization. One of the measurable effects is that the phase difference between the ingoing wave from the source and the reflected wave is multi-valued.
The properties of this system will be investigated by a simpler set-up, a multi-layer structure (cf. Figure 2). This makes the analysis easier and maintains the essential features of course. We consider normally incident scalar waves, and assume that the fields do not depend on the coordinates parallel to the layers. Such waves are described by the Helmholtz equation with -for a Kerr material- a cubic nonlinearity. The equation can be seen as that of a nonlinear oscillator, with the distance z along the grating as the evolution parameter. The refractive index depends on z, and the frequency of the incoming light comes in as a parameter. The equation is Hamiltonian, which simplifies the interpretation and analysis of the solutions.

Multi-valuedness in a grating was analyzed for the first time in [1], on the basis of a coupled mode model, and in [2] and [3], where numerical calculations were presented for a finite multi-layer. These papers are devoted primarily to the transmission problem, including soliton propagation. The same holds for most of the experimental papers. A relatively recent survey of experimental and theoretical work on a grating is found in [4]; see also [5]. The authors are not aware of experiments where the zero flux situation is guaranteed.
Standing waves, in particular their bifurcation from a homogeneous system, are discussed in [6]. These authors derive an autonomous Hamiltonian system for the envelope fields. The corresponding phase portraits describe essentially the structure of the phase portrait of the problem considered here, but are valid for a restricted set of parameters only, and 'chaotic' behavior is ruled out.
More generally, the present problem can be described within the coupled mode approach. An early account where standing waves were adressed explicitly is given in

[7] Many papers on coupled mode method have been written, c.f. [8] and recently [9]. Applying this method to standing waves, one again finds an autonomous Hamiltonian system, as in [6].
Numerical and analytical investigations in the spirit of the present work, but on a rather simplified model for the grating, are reported in [10], [11] and [12].

2. The equations of motion

Consider a scalar field in the y -direction say that propagates in the z -direction. Assume continuous waves

$$E(z,t) = E(z)e^{-i\omega t} \tag{1}$$

where E is a complex amplitude. Then E satisfies the scalar Helmholtz equation

$$\frac{d^2E}{dz^2} + k^2E = -k^2\gamma|E|^2 E \qquad \text{where } k = \frac{\omega n}{c} \text{ and } \gamma = \frac{3\chi^{(3)}}{n^2} \tag{2}$$

Our interest is in standing waves, i.e. E is real. Then (2) can be seen as the equation of a real oscillator of one degree of freedom in a quartic potential, with z as the evolution parameter. Spatial inhomogeneity enters this equation via $n = n(z)$ and $\chi^{(3)}(z)$.

These equations are applied on a system of bi-layers as shown in Figure 2. In our calculations the medium a is nonlinear, medium b is linear with respective linear refraction indices n_a and n_b. Within each layer the equations are independent of z. Between the layers one requires continuity of E and the spatial derivative E_z.

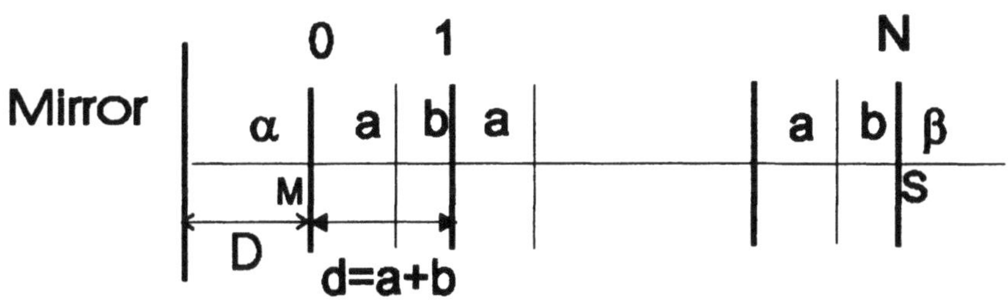

Figure 2: One realization of the experimental set-up: a multi-layered structure.

2.1 AN APPROXIMATE EQUATION FOR A HOMOGENEOUS LAYER

Recall that we consider real E. The following transformation, where E_z denotes dE/dz

$$E=\frac{1}{\sqrt{2k}}(Q+Q^{*}) \quad , \quad E_{z}=i\sqrt{\frac{k}{2}}(-Q+Q^{*}) \tag{3}$$

with inverse,

$$Q=\sqrt{\frac{k}{2}}(E+\frac{i}{k}E_{z}) \quad , \quad Q^{*}=\sqrt{\frac{k}{2}}(E-\frac{i}{k}E_{z}) \tag{4}$$

yields a first order equation for Q, which (together with its complex conjugate) is equivalent to (2)

$$d_{z}Q+ikQ=-i\frac{\gamma}{4}\left|Q+Q^{*}\right|^{2}(Q+Q^{*}) \tag{5}$$

The nonlinear term is not preceded by a k anymore, since the wavenumber k has been incorporated in Q.

For a linear material γ equals zero. The solution is $Q(z)=Q(0)\exp{-ikz}$, and the solution moves with constant phase speed k on a circle in the complex Q-plane. When the nonlinearity is included, the trajectory is on an ellipse that deviates slightly from the circle, and the phase velocity is shifted and varies along the orbit. Neglecting the slight deviation from a circle and the variation of the phase velocity, one can simplify (5) by taking into account only terms $\left|Q\right|^{2}Q$ in the r.h.s. Then one is left with

$$d_{z}Q+ikQ=-i\frac{3}{4}\gamma\,|Q|^{2}\,Q \tag{6}$$

For the present purpose this is justified, since the main effects are due to the switching between two different layer. As in the linear case the solutions form circles

$$Q(z)=Q_{0}\exp(i(k+\frac{3}{4}\gamma\,|Q_{0}|^{2})z) \tag{7}$$

and have constant phase velocity, which is shifted with respect to k. In this way, one neglects the variations in phase velocity and deviations of the circular shape. One expects that these two effects influence the behavior we are interested in only quantitatively, and even in that respect only to a small extent. We will use primarily (6), rather than (5).

One way to interprete Q is as follows. For any solution in the grating, Q can be written as a Fourier decomposition $Q=\int dkq(k)\exp(ikz)$. Thus from (3) one sees that E is the sum of two infinite sets of traveling waves running in opposite direction. Thus for the specific case that there is only one k involved, one interpretes the phase of Q is as (half of) the phase difference between the two in opposite direction running waves. The energy flow in each of these two waves is determined by the Poynting vector $\vec{E}_{real}\times\vec{H}_{real}$. Evaluating this expression and averaging over one time period of the motion one finds for the energy flow in each of the traveling waves

$$P = \frac{1}{4}\mu_0^{-1}\omega^{-1} |Q|^2 = \frac{1}{4}c\varepsilon_0 nk^{-1} |Q|^2 \tag{8}$$

Consequently, $|Q|^2$ determines the energy flow of a traveling wave in a linear medium. This applies in particular to the intensity of the source. Furthermore with $|E_z| = \omega\mu_0 |H|$ one sees that

$$|Q|^2 = \frac{k}{2}(E^2 + \frac{1}{2k}E_z^2) \tag{9}$$

is proportional to the local energy density.

2.2 CONTROL PARAMETERS $|Q_{SOURCE}|$ AND ϕ_{MIRROR}.

In the given experimental set-up, one controls the amplitude of the incoming wave at the source side, and via the distance D, the phase of Q at the mirror side. These two variables determine the trajectory inside the grating. In detail this works as follows. At the mirror the electric field is zero. Consequently, through (4), Q at the mirror has the form $Q_0 = i|Q_0|$. Then at the beginning of the grating the field is given by

$$Q_m = i|Q_0|e^{ik_a D} \tag{10}$$

where k_a is the wave number (linear) medium between grating and mirror. The connection with the field at the source side is then given by

$$Q_S = M_N(Q_M) \tag{11}$$

where M_N is the nonlinear transfer map between S and M.

This relation gives the field at the *source* side in terms of the field at the *mirror* side. Note that there is a one-one relationship between Q_S and Q_M. In the experiment, however, we control the amplitude of the source, $|Q_S|$ and the phase arg Q_M at the mirror side. When the system is linear these two real quantities determine uniquely the field in the grating. For a nonlinear system the fields are not uniquely determined by these two quantities as we will see.

2.3 THE TRANSFERMAP FOR THE GRATING

The transfer map M_N is determined by the so-called Poincaré map, that gives the transfer map for one unit of the grating. Let Q' be the field at the beginning of a layer a, the field Q'' at the beginning of the next layer a is given by the composed transformation

$$Q'' = T(Q') \qquad \text{where} \qquad T = B_{ba} \circ \Phi_b \circ B_{ab} \circ \Phi_a \tag{12}$$

Here Φ_a is the flow through the layer a and B_{ab} is a linear transformation of Q, taking care of the boundary conditions going from a to b. It holds

$$M_N = B_{b\beta} \circ T \circ T \circ .. \circ T \circ B_{\alpha a} = T^N \tag{13}$$

The transformation B_{ab} between two materials is determined by the requirement of continuity of E and E_z. It can be represented by a 2×2 matrix acting on a pair $\mathrm{col}[Q, Q^*]$. With the transformation in (4) one finds

$$B_{ab} = \begin{bmatrix} \beta_+ & \beta_- \\ \beta_- & \beta_+ \end{bmatrix} \text{ where } \beta_\pm = \frac{1}{2}(\rho \pm \rho^{-1}), \quad \rho = \sqrt{\frac{k_b}{k_a}} = \sqrt{\frac{n_b}{n_a}} \tag{14}$$

For the linear material b the flow trivially is given by the matrix

$$\Phi_b = \begin{bmatrix} e^{-ik_b b} & 0 \\ 0 & e^{ik_b b} \end{bmatrix} \qquad \text{with} \qquad k_b = \omega c^{-1} n_b \tag{15}$$

For a nonlinear system Φ can generally not be expessed in terms of a matrix. However, with the solution (7) of the approximate equation (6) the transfermap in the nonlinear medium a can be given in the form of a matrix

$$\Phi_a(Q) = \begin{bmatrix} e^{-ik_{nl} a} & 0 \\ 0 & e^{ik_{nl} a} \end{bmatrix} \begin{bmatrix} Q \\ Q^* \end{bmatrix} \qquad \text{with} \qquad k_{nl} = k_a + \frac{3\gamma}{4}|Q|^2 \tag{16}$$

2.4 PERIODIC SOLUTIONS IN THE INFINITE GRATING

Periodic orbits are of great importance for understanding the behavior of the system. Consider the infinte layered system with equal refraction indices $n_a = n_b$. Consequently it holds $B_{ab} = B_{ba} = Id$. Then the transfer map leaves $|Q|$ unchanged and the phase of Q increases with an amount

$$k_a a + k_b b + \frac{3}{4}\gamma |Q|^2 a \tag{17}$$

as one may verify. In particular this phase equals ±π, and the mapping has period-2 and period-1 trajectories when

$$\frac{3}{4}\gamma\,|Q|^2\,a = m\pi - \kappa n_{av} \text{ where } \kappa = k_0 d \text{ and } n_{av} = n_a\frac{a}{d} + n_b\frac{b}{d} \tag{18}$$

Here k_0 is the wavelength in vacuum.

Now make the indices of refraction different. Then band gaps in the linear system originate, and bifurcations in the nonlinear system may occur precisely at the values for the amplitude and the frequency that satisfy this relation. The reason is that only then the eigenvalues of the mapping are degenerate, which (for the present Hamiltonian system) is a prerequisite for a bifurcation.
Indeed, for $|Q|=0$ one is left with the the condition on the dimensionless quantity κ for which a band gap will arise, and splitting of the periodic orbits will be observed for nonzero $|Q|$.

3. Grating with linear response

3.1 EIGENVALUES, TRAJECTORIES AND BAND GAPS

In the case of material with a linear response the mapping (12) is linear. For the analysis it is useful to use a real representation for Q so that the mapping is given by

$$V_{n+1} = T_R V_n \quad \text{where} \quad V = \begin{bmatrix} X \\ Y \end{bmatrix} \quad \text{and} \quad X + iY = Q \tag{19}$$

where T_R is a real matrix. Eigenvalues $\lambda_\pm$ and eigenvectors $A_\pm$ of T_R determine the trajectories in a system with N units. These are then explicitly given by

$$V_n = \lambda_+^n A_+ \alpha_+ + \lambda_-^n A_- \alpha_- \tag{20}$$

where $\alpha_\pm$ is determined by the 'initial condition' V_0. Since det $T_R = 1$ the eigenvalues satisfy $\lambda_+\lambda_- = 1$ and they are either on the unit circle, or on the real axis. In the latter case one defines $|\lambda_+| > 1$ and $|\lambda_-| < 1$.

As a function of $\kappa = k_0 d$ the absolute value of the eigenvalues typically behave as shown in Figure 3. In the first gap, width Δ_-, both eigenvalues are negative. In the second one, width Δ_+, the eigenvalues are positive. Note the large difference in the two cases, $n_a = 3, n_b = 1$ and $n_a = 2.1, n_b = 1.9$ respectively.

When the eigenvalues are complex, the trajectories oscillate with n and remain bounded. For real eigenvalues a trajectory contains typically a decaying and a growing component. We are primarily interested in the latter case.

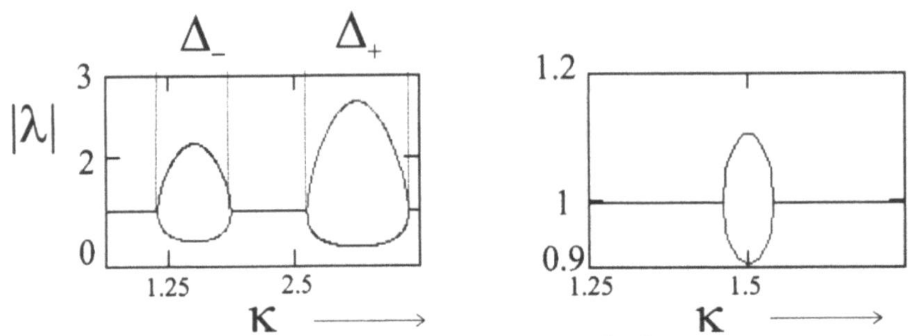

Figure 3: The absolute value of the eigenvalues as function of $k_0 d$. for $n_a = 3, n_b = 1$ and $n_a = 2.1, n_b = 1.9$ respectively . $\Delta_{-(+)}$ indicates the first and second band gap respectively, where the eigenvalues are negative (positive).

3.2 FIELDS AS DETERMINED BY ϕ_{MIRROR} AND $|Q_{SOURCE}|$

For a linear system, the field in the grating is determined uniquely by ϕ_{MIRROR} and $|Q_{SOURCE}|$ as one expects.

For a fixed intensity of the source, the possible field Q_S are on a circle $(Q_S, Q_S) = |Q_S|^2$ in the complex Q-plane, or equivalently the $X - Y$ plane. Then Q_M satisfies

$$(Q_M, M_N^\dagger M_N Q_M) | Q_S |^2 \tag{21}$$

As a consequence the possible values of Q_M are on an ellipse, which is determined by the symmetric poitive matrix $M_N^\dagger M_N$. Note that the area of the ellipse is equal to that of the circle, since $\det M_N^\dagger M_N = 1$.

The orientation of the ellipse is given by the eigenvectors of $M_N^\dagger M_N$ and the ratio of the axes of this ellipse is determined by its eigenvalues $\mu_+ > \mu_-$. It holds $axis_{long} / axis_{short} = \mu_+ / \mu_-$. In particular, when T_R has real eigenvalues, one expects that the ellipse is long, as drawn.

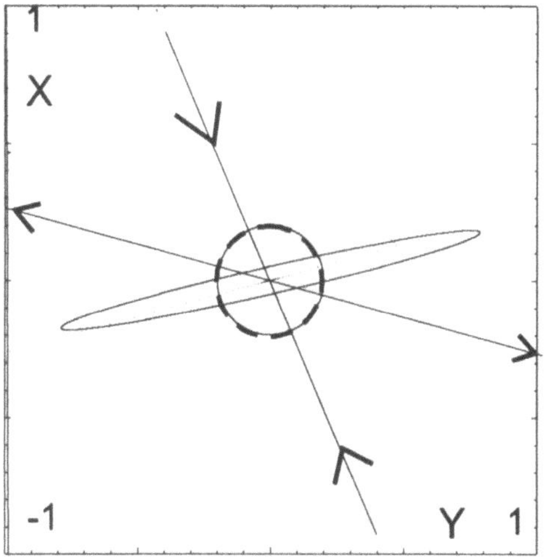

Figure 4: Deformation of a circle to an ellipse as the effect of T^N. Parameters are: $n_a = 2.1, n_b = 1.9, \gamma = 0, |Q_S| = 0.2$. The circle has been iterated 15 times.

For a given $|Q_S|$, the phase ϕ_M determines $|Q_M|$ uniquely because of the ellipse shape. Obviously $|Q_M|$ is also linear expression in $|Q_S|$. Thus the amplitude at the mirror side is a function of $|Q_S|$ and ϕ_M of the form

$$|Q_M| = g(\phi_M)|Q_S| \tag{22}$$

The situation is illustrated in Figure 4. Note that the transformation $Q_S \to Q_M$ is invertible, so that each point on the ellipse corresponds to one point on the circle, and vice versa. Note in particular that the ellipse covers the so-called unstable manifold. This is the locus of trajectories of the special solutions $V_n = \lambda_+^n A_+ \alpha_+$. From the construction in the graph one sees that $g(\phi_M)$ is peaked around the long axis of the ellipse.

Consider the behavior of this ellipse for increasing N. If T_R has *complex* eigenvalues, the ellipse will vary with N, but will stay bounded in N, just as the trajectories given in (20) remain bounded. In contrast, if T_R has *real* eigenvalues, these trajectories grow with N. Correspondingly, for large N the eigenvalues $\mu_\pm$ behave as $\mu_+ \sim \lambda_+^N$ and $\mu_- = \mu_+^{-1} \sim \lambda_+^{-N}$. The direction of the long axis will converge to the direction of the eigen vector belonging to λ_+, and the ellipse becomes longer and thinner with N. Correspondingly the width of the peak of g is approximately equal to the ratio of the axis of the ellipse. In fact one can show that for large N it holds approximately

$$g_{\max} \sim \lambda_+^N \qquad \text{and} \qquad g_{width} \sim \lambda_-^N = \lambda_+^{-N} \tag{23}$$

For the appreciation of the results in the nonlinear system one should note that the ellipse contains part of the unstable manifold as shown.

Field inside the grating

One observes that for linear gratings and frequency in the band gap, there is a small range of ϕ_M where $g_{\max} >> 1$. Then the field incoming from the source builds up to a high value at the mirror side. Outside this range, however, one sees the typical reflecting behavior: fields do not penetrate into the grating. Obviously, these effects are most pronounced in the center of the band gap, where λ_+ attains its maximum. Keeping N fixed, they disappear for frequencies coming closer to the edges of the band gap.

4. Grating with nonlinear response

Here we present results that are obtained on the basis of the mapping (12) with the approximate nonlinear flow in (16). We consider primarily frequencies near the upper edge of the band gap.

4.1 BIFURCATION OF PERIODIC TRAJECTORIES

When the frequency, i.e. k_0, varies, the eigenvalues of T_R move from the complex unit circle to the real axis and back. Correspondingly the zero-field state, i.e. the origin in the Q plane, changes from elliptic to hyperbolic and back. Because of the nonlinearity generically for increasing nonlinear parameter a periodic orbit bifurcates from the origin at each of the edges of the band gap. The period of these orbits will be 1 or 2 with respect to the period of the grating when the real eigenvalues $\lambda_\pm$ are positive or negative respectively. Thus, for the first band gap the period is 2, and it is 1 in the second. Similarly, for a given non-zero nonlinear parameter, one expects that at $|Q|$-values that satisfy (18), the periodic orbit that exists for $n_a = n_b$ will split in two orbits for increasing contrast of the index of refraction.
For fixed non-zero value of γ and non-zero index contrast we calculated the periodic trajectories and found curves as shown in Figure 5. Observe that the calculated curves are closer to one another for smaller index contrast. Indeed, they converge to the straight lines representing the periodic trajectories of the system with zero index contrast shown in the first graph. For the interpretation of the phase portraits below it is important to see that the slope of the bifurcating orbits is of the same order of magnitude as the slope of the lines of the zero contrast trajectory. The deviation depends on the grating parameters.

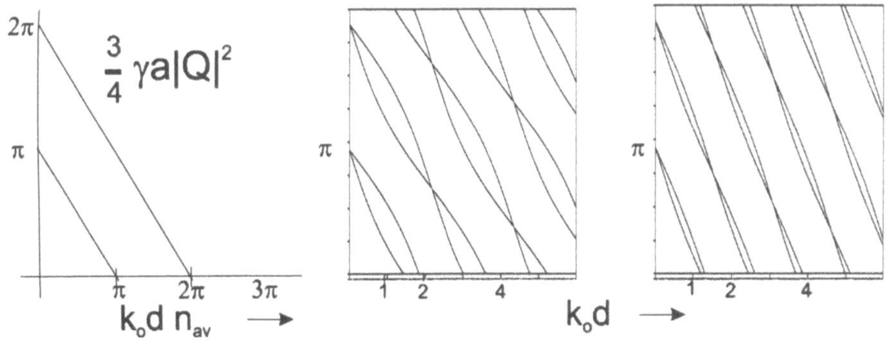

Figure 5: Left: straight lines where the periodic trajectories of the transfer map of the system with $n_a = n_b$ has period 2 or 1 (Section 2.4). Middle and right: calculated periodic trajectories. Parameters: $a/d = 0.15, n_a = 3$; left: $n_b = 1.7$, right: $n_b = 2.4$.

4.2 BOUNDEDNESS AND INSTABILITY OF THE ORIGIN

In Figure 6 a sequence of phase portraits is shown for decreasing values of $\kappa = k_0 d$. The axes in the graphs are respectively X and Y. Values of parameters in the calculation are chosen so that in terms of the electric field (at the beginning of layer a one should identify

$$X = \frac{3}{2\sqrt{2}}(k_a d)^{1/2}\sqrt{\chi^{(3)}}E$$
$$Y = \frac{3}{2\sqrt{2}}\frac{k_0 d}{(k_a d)^{1/2}}\sqrt{\chi^{(3)}}k_0^{-1}E_z \tag{24}$$

For the present purpose the main features of the phase portraits are

- The closed curves that enclose the origin;
- For decreasing κ the origin changes from elliptic to hyperbolic, and at the same time a 'figure 8' structure develops.

These are phase portraits of a Hamiltonian system with a periodic perturbation. For general aspects of interpretation there are many suitable texts on 'Nonlinear Dynamics'. For a simple introduction see e.g. [13]. Several topics are also discussed in [14]. The phase portraits are calculated with the numerical programme that goes with this book. The closed curves around the 'figure 8' represent trajectories that stay on such a curve for any number of units in the grating: they go neither to zero, nor to infinity. Furthermore one should realize that a trajectory in between closed curves, stays there. Consequently the fields Q_n in between Q_S and Q_M behave regular and each trajectory has both a lower bound and an upper bound.

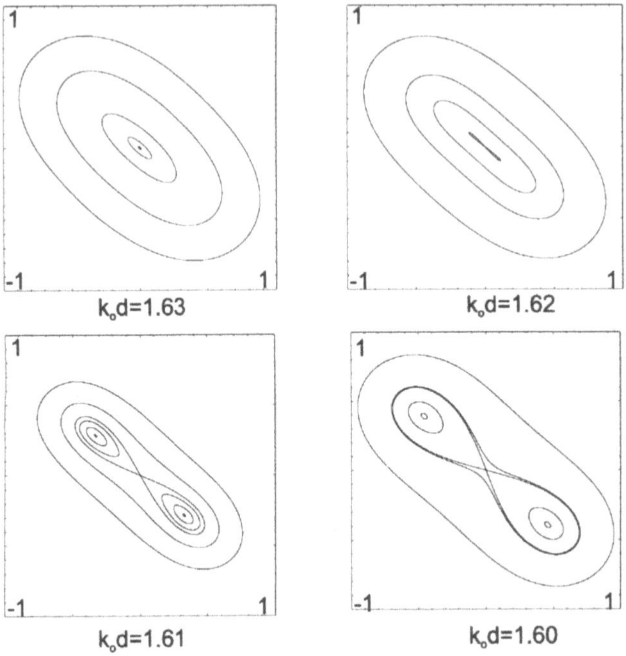

Figure 6: A sequence of phase portraits for decreasing values of $k_0 d$: 1.63, 1.62, 1.61, 1.60 respectively. System parameters are: $a/d = 0.5, n_a = 2.1, n_b = 1.9$.

The 'figure 8' represents a so-called 'homoclinic trajectory', that connects the origin with itself. An initial condition starting on the '8' moves forward towards the origin, for $n \to \infty$, and comes from the origin for $n \to -\infty$. Here we are dealing with the first band gap, where the eigenvalue of the origin is negative, so that the trajectory swaps between the two lobes of the '8'.

In the '8' one sees invariant circles, enclosing the period 2 trajectory that has bifurcated from the origin. The size of this 'figure 8' is of the order of the distance of the equilibrium trajectory in its centers. As seen in Figure 5 it is roughly given by the slope of the periodic trajectories in the homogeneous material

$$|Q|^2 \approx |\kappa_u - \kappa| \, n_{av} \frac{3}{4a} \gamma_a^{-1} \tag{25}$$

where κ_u corresponds to the upper bound of the band gap. In terms of the electric field one finds, cf. (4) and (2)

$$\frac{k_a}{2}(E^2 + k_a^{-2} E_z^2) \approx |\kappa_u - \kappa| \frac{1}{a\chi^{(3)}} n_a^2 n_{av} \tag{26}$$

Remark that g is determined by the geometry and the linear optical properties,

One conclusion that can be drawn from these considerations is the following: the maximum growth of a *small* initial condition $|Q_S|$ is of the order of the r.h.s in (26). This maximum growth is independent of the number of units N. This is in contrast with the situation in the linear case, where for a suitable phase at the mirror side, the trajectories may walk of to infinity with N. Thus the nonlinearity yields a bound for the growth of a small input intensity from the source.

4.3 MULTI-VALUEDNESS

Multi-valuedness arises as follows: as in the linear case we consider the effect of T^n on a (small) circle of initial conditions Q_S.

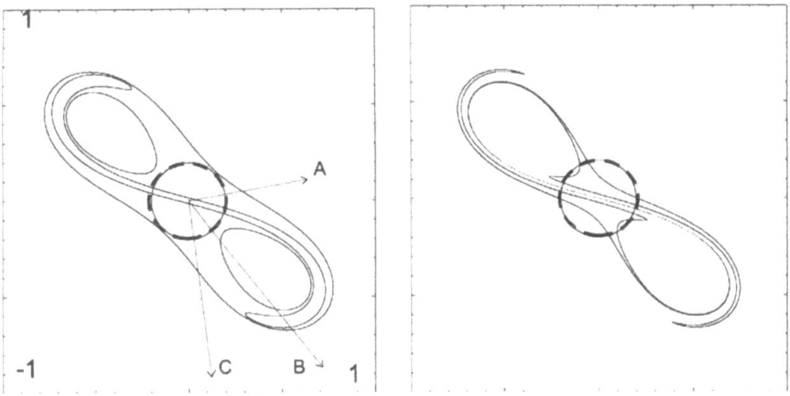

Figure 7: Iterated mapping of a circle $|Q_S| = 0.2$ showing the origin of multi-valuedness; $k_0 d = 1.60$. Left: the circle is mapped after 30 times to an S-shaped area. This area is enclosed within the outer closed invariant curve and the two inner closed invariant curves. Right: the circle after 75 iterates.

A situation as sketched in Figure 7 can be obtained for a proper choice of N and the eigenvalue $|\lambda_+| > 1$ in the band gap (cf. Figure 3). For sufficiently large λ_+^N, a circle of initial conditions is deformed into a long thin shape, with the same area as the original circle, that stretches along the unstable direction of the origin. The exact shape depends of course on N and on $|Q_S|$.

Multi-valuedness arises since there is a range of ϕ_M where one value of ϕ_M leads to more than one value of Q_M. Each point $[\phi_M, |Q_M|]$ is connected to a different point on the circle $|Q_S|$. Consequently, for each ϕ_M in this range there are different trajectories connecting the source- and the mirror side.

In the example shown the multi-valuedness equals three, for the direction B in the graph. This is a situation in which typically hysteresis can be observed: Suppose one starts with ϕ_M in the direction A, where the solution is single-valued. When ϕ_M is increased slowly to B, the solution is expected to follow the upper branch of the S-shape. After B, where the solution becomes single-valued, the solution 'falls down' to a

lower value of $|Q_M|$. Then decreasing ϕ_M to its original value one expects to be able to create a hystersis loop.
For larger N and the other parameters fixed, the circle in Figure 7 deformes to a much longer and thinner band that follows the unstable manifold, as in the right graph. This leads to *bands* of values of $|Q_M|$ with a complicated internal structure. These bands remain contained in the area between the closed invariant circles in the left graph.
Clearly what actually is observed in an experiment, depends on the parameters κ and ϕ_M and $|Q_S|$, where κ comes into play via $\lambda_+(\kappa)^N$. In particular, keeping the first two fixed one observes linear single-valued behavior for $|Q_S|$ 'small enough'.
However, for κ further away from the edge of the band gap, λ_+^N becomes bigger and the multi-valuedness appears.
One interesting effect is that due to the nonlinearity the maximum field intensities may be large, but remain bounded (cf. 26) and do not depend on N. The region of ϕ_M in which these values occur is determined by the phase portrait, and equals the angle between the stable and the unstable direction of the origin (cf. Figure 6). From the graphs we see that these are an appreciable part of 2π. This is in contrast with the linear case where extremely *high* fields of the order λ_+^N can occur at the mirror side, however for an extremely *small* region of ϕ_M.

4.4 UNBOUNDEDNESS AND 'ERRATIC' TRAJECTORIES

In the graphs in Figure 6 one sees that there is a 'largest' closed curve around the 'figure 8'. Outside that curve the trajectories are unbounded. This unboundedness extends to the origin for larger index contrast as one sees in Figure 8.
In this graph one sees at $\kappa = 1.85$ no essential difference with $\kappa = 1.60$ above. For $\kappa = 1.80$ however one observes a thickening of the 'figure 8'. This is caused by the effect of homoclinic tangles: these are present in all cases, but much more dominant here. For $\kappa = 1.75$ the 'figure 8' has almost disappeared and the closed curves are absent. The dots are elements of unbounded trajectories.
In the last graph the image of a circle of radius 0.3 is shown after 9 iterates of the mapping. This image stretches along the unstable manifold, thus its shape is caused by the homoclinic tangles of the latter.
A first consequence is that the multi-valuedness is no longer organized in bands as above. In particular, because of the absence of the outer closed curves, there is no upper bound on the multi-valued trajectories.
In this graph the radius of the circle of initial conditions is unrealistically large.
However, similar things happen for arbitrarily small $|Q_S|$ if N is sufficiently large.
Consequently the conclusion hold true for more realistic values of $|Q_S|$.

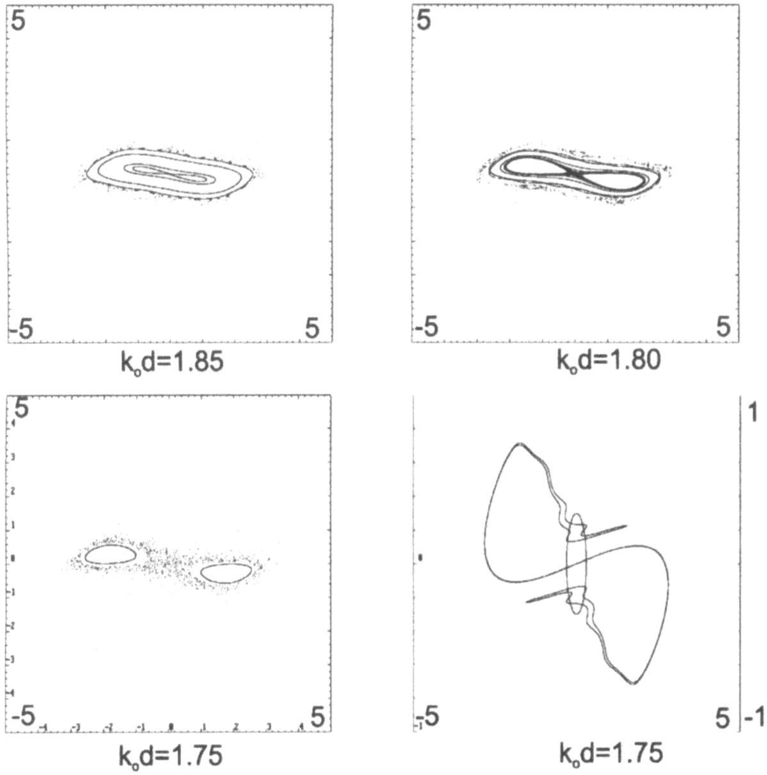

Figure 8: A sequence of phase portraits for decreasing values of $\kappa = k_0 d$. System parameters are $a/d = 0.5$, $n_a = 3$ and $n_b = 1$. In the last graph one sees a circle of radius 0.3 and its image after 9 iterates (note the different y-scale)

5. DISCUSSION

In this paper we dicussed the consequence of nonlinearity and periodicity for continuous standing waves in the proposed structure. It was shown that multi-valuedness and erratic behavior can occur for realistic systems. This was done for a relatively simple model, of which we claim however that it contains the essential features of any periodic system with a bandgap.

The multi-valuedness results are also found in approximate models, such as from coupled mode theory or bifurcation analysis. These models, however, are derived under certain restrictions, primarily small index contrast and slow change of the field envelopes with respect to the grating period Because of the absence of these restrictions, the results of the present system have a more general applicability. In particular it allows the investigation of the effects like homoclinic tangles and break-up of tori, leading to unbounded trajectories.

The dynamic stability of the solutions found falls outside the scope of this paper and should be subject of further study. In the present set up, this stability has to be investigated having in mind that ϕ_M and $| Q_S |$ are the physical control parameters. This leads to a specific formulation of the stability problem.

In the model absorption was neglected. When this is included, one expects attracting periodic trajectories and possibly spatial strange attractors

References

[1] Winful, H.G., Marburger, J.H. and Garmire, E, Theory of bistability in nonlinear distributed feedback structures, *Appl. Phys. Lett. 35*, 379-381 (1979)

[2] Chen, W. and Mills D.L., Gap solitons and the nonlinear optical response of superlattices, *Phys. Rev. Lett. 58*, 160-163 (1987).

[3] Chen, W. and Mills D.L., Optical response of nonlinear multilayer structures: bilayers and superlattices, *Phys. Rev. B 36*, 6269-6278 (1987).

[4] Focus Issue: Bragg Solitons and Nonlinear Optics of Periodic Structures, *Optics Express, 3* (1998).

[5] Eggleton B.J., Martijn de Sterke C. and Slusher R.E., Bragg solitons in the nonlinear Schrödinger limit: experiment and theory, *J. Opt. Soc. B 4* , 587-599 (1999).

[6] Kovalev A.S., Usatenko O.V. and Gorbatch A.V., Bifurcation picture for gap solitons in nonlinear modulated systems, *Phys. Rev. E 60*, 2309-2316 (1999).

[7] Mills, D.L. and Trullinger, S.E. Gap solitons in nonlinear periodic structures, *Phys. Rev. B. 36*, 947-952 (1987).

[8] De Sterke, C.M., Salinas, D.G. and Sipe, J.E., Coupled-mode theory for light propagation through deep nonlinear gratings, *Phys. Rev. E 54*, 1969-1989 (1996).

[9] Iizuka, T. and De Sterke, C.M. Corrections to coupled mode theory for deep gratings, *Phys. Rev. E 61*, 4491-4499 (2000).

[10] Coste, J. and Peyraud, J., Stationary waves in a nonlinear periodic medium: Strong resonances and localized structures. I. The discrete model, *Phys. Rev. B 39*, 13086-13095 (1989).

[11] Coste, J. and Peyraud, J., Stationary waves in a nonlinear periodic medium: Strong resonances and localized structures. II. The continuous model, *Phys. Rev. B 39*, 13096-13105 (1989).

[12] Peyraud, J. and Coste, J., Dynamics of gap solitons, *Phys. Rev. B 40*, 12201-12208 (1989).

[13] Hilborn R.C., Chaos and Nonlinear Dynamics, *Oxford University Press*, Oxford (1994).

[14] Nusse, H.E. and Yorke, J.A. *Dynamics: Numerical Explorations* (Springer-Verlag New York Inc., New York, 1998). The book is accompanied by the computer program *Dynamics 2*, coauthored by Hunt, B.R. and Kostelich, E.J.

STORAGE DENSITY ENHANCEMENT IN HOLOGRAPHIC MEMORY CARD SYSTEM

P. KOPPA[*a], P. VÁRHEGYI[b], T. UJVÁRI[b], G. SZARVAS[b] S. HVILSTED[c], P. S. RAMANUJAM[c], P. RICHTER[a]
[a] *Technical University of Budapest, Dept. of Atomic Physics, Budafoki ut 8, 1111 Budapest, Hungary,*
[b] *Optilink Ltd, Graphisoft Park, 1031 Budapest, Hungary*
[c] *Risø National Laboratory, DK-4000 Roskilde Denmark*
[*] *Correspondence: E-mail: Koppa@eik.bme.hu, Telephone: 36 1 463 39 66*

Key words:

Holographic memory, Fourier hologram, Storage density

Abstract:

Our goal is increase the storage density of our rewritable holographic memory card system based on thin film polymer media on credit card size plastic carriers. We use computer simulation to optimize the illumination and the Fourier filtering in order to have optimal trade-off between bit error rate and data density.We also investigate the possibility of using gray scale data coding.

Introduction

We are developing a new Holographic Memory Card (HMC) system that provides the advantages of optical storage in a card format in order to largely exceed the performances of present cards. Compared to bit-serial optical data storage (employed in CD, DVD and existing optical cards), our page organized holographic storage technology presents a number of advantages like higher immunity to local defects, parallel access read and write systems, lower card positioning and fabrication tolerances and no focus and track servo mechanisms are needed.
In this study we investigate the optimization of the design parameters for enhancing storage density. We developed a computer simulation model of the system, based on fast Fourier transformation. In the model we account for several important features and noise sources. With the model we evaluated the key performances of the holographic memory system, and investigated two main methods for increasing the storage density.

A. Driessen (ed.), Nonlinear Optics for the Information Society, 179–184.

These are gray-scale encoding and Fourier filtering. Both methods reduce the available Signal-to-Noise Ratio (SNR), so tradeoff between high capacity and low error rate is to be investigated.
Gray-scale encoding [3], [4], in which pixels take one of g brightness levels, affect both capacity and readout rate, because each pixel now conveys $\log_2 g$ bits of data. So the gray-scale encoding multiplies the storage density by $\log_2 g$, but divides the available SNR into g - 1 parts.
Fourier filtering means changing the aperture stop of the system [5]. It can be done without any other changes to the device. The aperture acts like a Fourier filter, so it sets the resolution and the size of the holograms. In this system holograms are stored side by side. So if we decrease the Fourier filter by the ratio of r, the hologram size also decreases, and the storage density multiplies by $(1 / r)^2$. However, the small aperture acts as spatially low-pass filter for the data-bearing object beam, and may decreases the SNR.

1. Areal storage density in the HMC system

The HMC system has some features that directly affect data density. These features have to be taken into account in the simulation and / or can be changed during the optimization of the system.

- If we use a low-pass filter, like the hologram aperture in our system, the high-spatial-frequency components of the data pattern displayed on the SLM will not be recorded by the holographic medium. As a result, the light propagated from a SLM pixel is spread to the other pixels. This source of errors is called interpixel cross talk [5]. As described in section 3, decreasing the hologram aperture can increase the storage density, but it also gives rise to interpixel cross talk.
- The SLM fill factor is not 100%. This means, that only a part of each SLM pixel transmits the laser light. This is an important factor, because with combining with the low-pass filter it is the main cause of the interpixel cross talk, described above.
- The shape of the reference and reconstructing beam can be plane wave or Gaussian beams. This can change the characteristics of the hologram and the reconstructed image.
- In this system we use a random phase mask. It spatially changes the phase of the wave front. Because of these changes the intensity distribution of the Fourier plane is smoother, the very high central peak of the Fourier plane is spread around. But it can also cause some interference in the image.
- In our system the SLM and the CCD are not pixel matched. To decode the data from the image, for every SLM pixel there must be a CCD pixel, which only receives light from that SLM pixel. This means that SLM pixels must be at least 2 times bigger, than the CCD pixels. This minimal oversampling can also be derived from Shannon's theory.
- In the SLM there can be difference between pixels. It has a deterministic part, when pixels are different spatially, and it can also change randomly in time. But the

deterministic part is also different from one SLM to another (from one HMC device to another), so the SLM noise can be modeled by random noise.

- The SLM has a finite contrast. This means, that the dark pixels are not true black, they transmit some light.
- The CCD also may have different pixels, and there is a random noise on each pixel and noise in the electronics of the CCD. These noises together can be modeled by Gaussian random noise.
- The hologram material causes random Gaussian noise.

When we take all these features and noises into account, the system output becomes similar to Fig 1. This image is sampled according to the pitch of the CCD, created with calculating the spatial average intensity to each CCD pixels. During decoding we have to find the CCD pixels that belong to single SLM pixels. Fig 1 also shows the histogram of these CCD pixels. It shows that the OFF and the ON pixels overlap, we have errors during decoding. The raw BER was 0.02 that is not acceptable. In a practical system raw BER should be less than 10^{-3}, that can be handled by error correction codes.

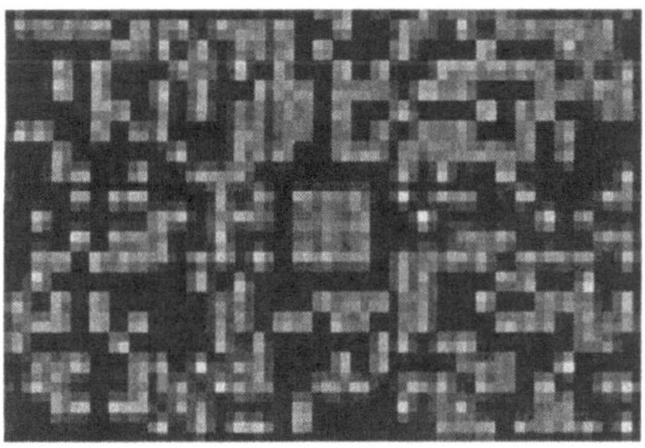
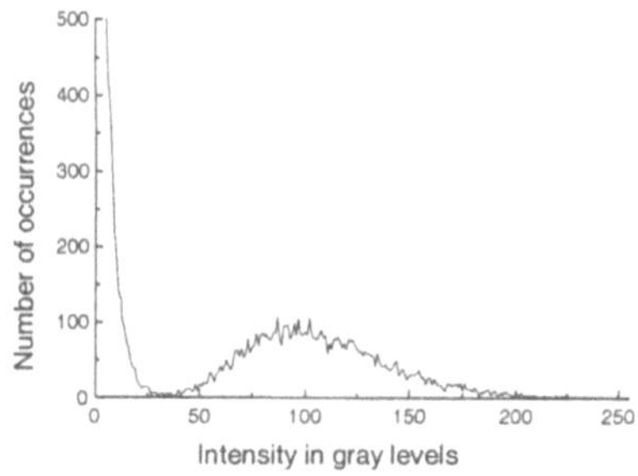

Figure 1: Simulation of the real system.

The optical design determined the SLM size and pitch size. We chose an SLM that fits to these parameters, but has a fill factor of 50% in horizontal direction and 75% in vertical direction. Because of the difficulties with the decoding algorithms, we needed more oversampling than the original 2. We chose a CCD with pitch size 3.2 times smaller than the SLM pitch size. The fill factor of the CCD is nearly 100%. With these fixed parameters we optimized the Fourier filter, the shape of the reference beam, the number of gray scales, and the random phase mask.

We investigated different types of reference beams, plane wave and Gaussian beams. The theory was that the SLM pixel's point spread function convolutes with the Gaussian beam, and results less interpixel cross talk. Our investigations proved this. Figure 2 shows the effect of the Gaussian beam. The X-axis is the $1 / e^2$ radius of the beam, and the Y-axis is the distance in gray levels between the brightest OFF and the darkest ON pixels on the simulated histograms. Positive distance means 0 BER and tolerance to noises. The optimum is the $1/e^2$ radius of 0.18 mm. The effect of the Fourier filtering (thus the hologram size) is shown on figure 2/b relative to the Nyquist aperture (0.29

mm with in the original setup.) The results show that at about 0.7 Nyquist aperture we still have some distance between brightest OFF and the darkest ON pixels and the hologram size is reduced by 0.7^2=0.49. This would mean a capacity growth of a factor of 2 with respect to the original setup. In this simulation we took into account the SLM's fill factor, but no other noises.

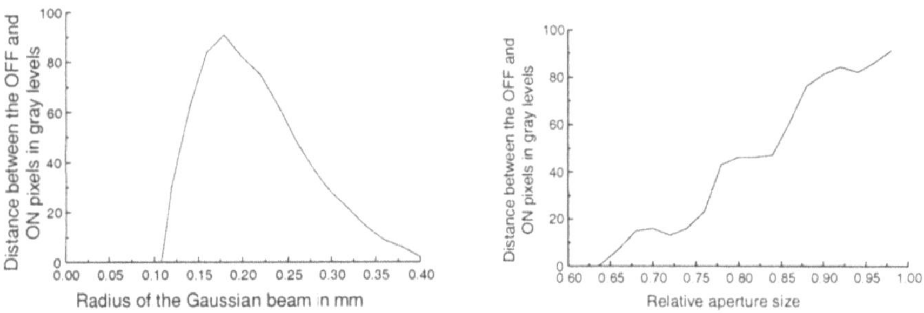

Figure 2: Optimization of the critical system parameters: Fourier filtering (a.) and Gaussian reference beam (b.)

We repeated the simulation with noises, and obtained a little different optimal Fourier filtering and reference beam. Figure 3 shows the image with this optimal Fourier filter and reference beam, and with the same noises as before. This means the storage density is 1.53 bit / μm^2. In the simulation the BER was $2*10^{-4}$.

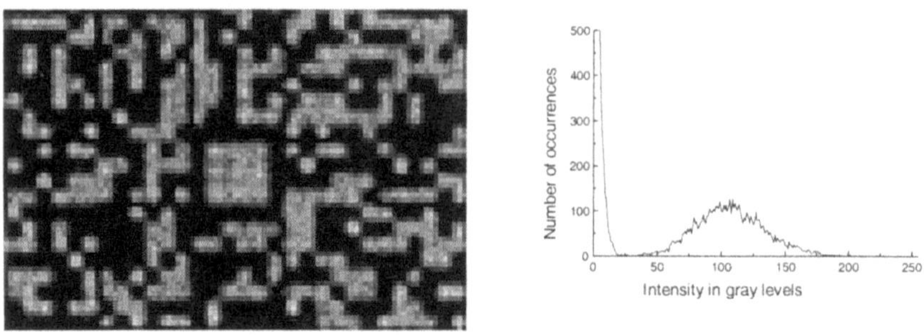

Figure 3: Simulation of the system with the optimal Gaussian reference beam and Fourier filter size

We have simulated the system with grey level encoding for up to 4 levels. Figure 4 shows the histogram of a 3 grey level image as an example. If using the optimal aperture and illumination, the three levels are overlapping, but the BER is only 10^{-3}. The storage density of this type of data is 1.12 bit / μm^2. We concluded that the multi-level coding is not a good alternative neither in data density nor in robustness against noises.

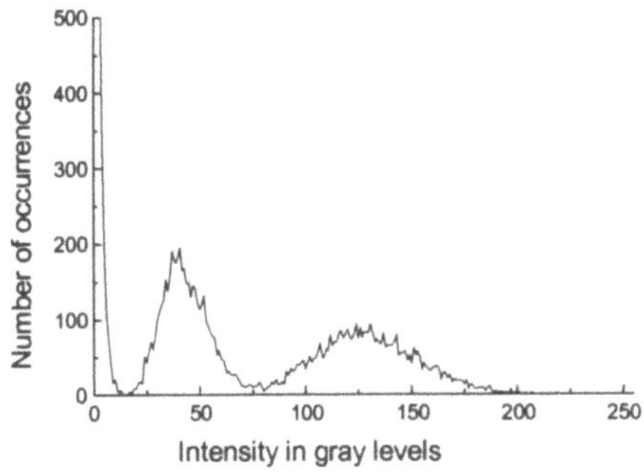

Figure 4: Histogram of a simulation with 3-brightness level.

We also tried other SLM fill factors, and the results showed that the system performance and the storage density could increase with increasing the SLM fill factor.

2. Conclusion

We have developed a model to optimize system performances, and evaluated parameter sets and methods increasing storage density. The model was used successfully to determine some of our holographic memory card drive's key parameters, and optimize its areal data storage density. We also investigated two main methods for increasing the storage density: the gray-scale encoding and the Fourier filtering.
The optimization has brought the system from a BER 0.02 and a data density of 0.75 bit / μm^2 to a maximum data density is 1.53 bit / μm^2 with the 0.7 relative aperture, Gaussian reference beam, and with 2 brightness levels. The BER was in this case $2*10^{-4}$. The multi-level coding can only be used with a lower storage density.

Acknowledgments

Parts of this work were supported by the Hungarian Academy of Science under the Bolyai grant.

References

[1] E. Lőrincz, F. Ujhelyi, A. Sütő, G. Szarvas, P. Koppa, G. Erdei, S. Hvilsted, P.S. Ramanujam, P.I. Richter: "Rewritable holographic memory card system", *Optical Data storage 2000*, Whistler, B. C. Canada, 15-17 May 2000, SPIE Vol. 4090

[2] P. Koppa, G. Erdei, F. Ujhelyi, P. Várhegyi, T. Ujvári, E. Lőrincz, G. Szarvas, S. Hvilsted, P. S. Ramanujam, P. Richter: "Data Storage on Holographic Memory Card", *Holography 2000*, 10-14 July 2000, St. Pölten, Austria.

[3] G. W. Burr, G. Barking, H. Coufal, J. A. Hoffnagle, C. M. Jefferson, and M. A. Neifeld: "Gray-scale data pages for digital holographic data storage", *Optics Letters, Vol. 23, No. 15*, pp. 1218-1220, 1998.

[4] S. Tao, B. Tang, Y. Zhou, L. Shen: "Quantitative study of the gray-scale fidelity of volume holographic images", *Applied Optics, Vol. 38, No. 17*, pp. 3767-3777, 1999.

[5] M-P. Bernal, G. W. Burr, H. Coufal, M. Quintailla: "Balancing interpixel cross talk and detector noise to optimize areal density in holographic storage systems", *Applied Optics, Vol. 37, No. 23*, pp. 5377-5385, 1998.

[6] J. W. Goodman: *"Introduction to Fourier Optics"*, McGraw-Hill Publishing Company, 1988.

[7] J. F. Heanue, M. C. Bashaw, L. Hesselink: "Channel codes for digital holographic data storage", *Journal of Optical Society of America A, Vol. 12, No. 11*, pp. 2432-2439, 1995.

[8] Á. Kerekes, Sz. Sajti, E. Lőrincz, S. Hvilsted, P. S. Ramanujam: "Rewritable azobenzene polyester for polarization holographic data storage", *Holography 2000*, 10-14 July 2000, St. Pölten, Austria.

SELF-ORGANIZED PARAMETRIC DOWN CONVERSION IN OPTICAL FIBRES

P. CHMELA
Institute of Physical Engineering, Brno University of Technology
Technická 2, 616 69 Brno, Czech Republic

Abstract

The theoretically predicted self-organized parametric down conversion in doped optical fibers was experimentally demonstrated for the first time and a weak permanent second-order susceptibility grating in the conditioned fiber was identified.

Since the discovery of self-organized frequency doubling in germanium-phosphorous-doped optical fibers by Österberg and Margulis [1] many ingenious experiments were performed and brilliant theoretical models were proposed (for review, see Ref. [2-6]), however, the phenomenon remains a puzzle hitherto.

The first puzzling question was how the initial second harmonic (SH) radiation can be generated in a centrosymetric optical fiber by the fundamental pump alone, without any external SH seed. Most physicists have believed that the initial self-seeded second-harmonic generation (SHG) occurs due to the quadrupole interactions [7]. However, other models have been suggested as well [2,8].

The essential question is what mechanism can break the centrosymmetric structure of the fiber or fiber preform being conditioned to SHG, and how the proper spatially periodic second-order susceptibility ($\chi^{(2)}$) structure can be created that enables an efficient quasi-phase-matched SHG.

It has been ascertained that a strong spatially periodic permanent dc electric field being of the order of some kV/cm is created at the interaction of fundamental and SH radiation in the SHG-conditioned samples [9-11]. This dc electric field deprives the centrosymetric structure of isotropic medium [12, 13] and enables an efficient quasi-phase-matched SHG [14-16].

The first phenomenological model was outlined by Stolen and Tom [17] who proposed that an internal dc electric field is produced owing to the non-degenerate third-order optical rectification,

$$\vec{\mathbf{P}}_{dc}^{(3)} = \frac{1}{2}\left[\chi^{(3)rect}(\omega+\omega-2\omega):\vec{\mathbf{E}}_{\omega}\vec{\mathbf{E}}_{\omega}\vec{\mathbf{E}}_{2\omega}^{*} + \mathrm{c.c.}\right] = \mathbf{e}_{dc}\chi_{eff}^{(3)rect}|A_{\omega}|^{2}|A_{2\omega}|\cos(\Delta kz - \Delta\varphi), \quad (1)$$

A. Driessen (ed.), Nonlinear Optics for the Information Society, 185–191.

where $\vec{E}_\omega = e_\omega A_\omega \exp[i(\omega t - k_\omega z)]$ and $\vec{E}_{2\omega} = e_{2\omega} A_{2\omega} \exp[i(2\omega t - k_{2\omega} z)]$ are the electric intensities of fundamental and SH light waves, $\Delta k = k_{2\omega} - 2k_\omega$ represents the phase-mismatch factor, $\Delta\varphi = \varphi_{2\omega} - 2\varphi_\omega$ is the relative phase shift between the interacting waves and $\chi^{(3)rect}(\omega + \omega - 2\omega)$ is the third-order rectification susceptibility.

The formation of a system of elementary electric dipoles was assumed and the internal dc electric field created, $\vec{E}_{dc} = -(\nu/\varepsilon_0)\vec{P}^{(3)}_{dc}$ (ν being a dimensionless constant and ε_0 is the permitivity of vacuum), was believed to deprivate the centrosymetric structure of the isotropic medium so that the SHG can occur due to the third-order nonlinear polarization [13]

$$\vec{P}^{(3)}_{2\omega} = \chi^{(3)dep}(\omega + \omega + 0) \vdots \vec{E}_\omega \vec{E}_\omega \vec{E}_{dc}. \tag{2}$$

In fact, the Stolen-Tom model describes the self-organized SHG as a fifth-order nonlinear optical interaction that cannot provide any efficient SHG in the instantaneous medium response picture. Though the $\chi^{(2)}$ grating created possesses the right spatial periodicity for an efficient quasi-phase-matched SHG [14-16], the SH radiation produced is shifted by $\pi/2$ with respect to the SH seed [5], which is precisely the wrong value to allow a weak SH seeding light to grow in the course of SHG preparation process.

The Stolen-Tom idea [17] gave rise to a group of orientational models [2, 18-21] that assume the formation of electric-dipole-related asymmetry or microscopic charge separation.

Another model explaining the $\chi^{(2)}$ grating formation by microscopic charge separation due to the photovoltaic effect was proposed by Dianov et al. [6, 22, 23]. An ionizing interference of two pump photons and one SH photon is assumed to cause the ejection of an electron from existing (photoinduced) defect center [21] in a preferred direction, which is trapped at the boundary of the illuminated area. The macroscopic charge redistribution induces then the spatially periodic $\chi^{(2)}$ grating. A photogalvanic space-oscillating coherent current j_{ph} is considered, which is a nonlinear function of the applied radiation field [6, 22, 23],

$$j_{ph} = \beta E^2_\omega E^*_{2\omega} + c.c. = |\beta| \, |A_\omega|^2 |A_{2\omega}| \cos(\Delta kz - \Delta\varphi + \psi), \tag{3}$$

where $|\beta|$ is the effective value of the coherent-photovoltaic tensor and ψ represents a phase constant shift [6]. As a response to the photoinduced dc electric field a conduction current $j_{ph} = \sigma E_{dc}$ occurs that works against the photogalvanic current.

Using the third Maxwell equation and the continuity equation it is easily to derive the equation for the time evolution of second-order susceptibility grating [6, 22, 23],

$$\frac{\partial\chi^{(2)}}{\partial t}+\frac{\sigma}{\varepsilon}\chi^{(2)}=\alpha A_{\omega}^{2}A_{2\omega}^{*}\exp[i(\Delta kz+\psi)]+c.c., \tag{4}$$

where ε is a permittivity and $\alpha=|\beta|\chi^{(3)dep}/\varepsilon$.

The macroscopic charge separation was identified in some bulk doped-glass samples being conditioned to SHG [6, 24].

Even when the photovoltaic model can explain many features of the behaviour of self-organized SHG [6] it shows also serious difficulties. In order to involve the freezing-in of the photoinduced $\chi^{(2)}$ grating it is necessary to assume that the conductivity σ drops immediately to zero after switching off the radiation field, which seems to be rather unnatural.

Moreover, Eq. (4) predicts that the saturation second-order susceptibility $\chi_{sat}^{(2)}$ depends on the instantaneous electric intensities of the saturation radiation field only, meaning that the resulting saturation state does not reflect the previous history of the preparation process, which is in strong disagreement with experimental observations (see, e.g. experiments by Kroll et al. [25] and their discussion [26]).

A directional photoionization model of self-organized SHG was elaborated by Anderson et al. [27] who considered two-, three- and four-photon ionizing interference to derive a new formula for steady-state dc electric field in the picture of anisotropic excitation process.

Though the Anderson's formula for steady state dc electric field, being formed in the course of self-organized SHG, was found to be in a strong disagreement with the experimental observations [28], the directional photoionization model yields good predictions for the saturation. It predicts for the saturation value of the ratio $I_{2\omega}/I_{\omega}^{2}$ (I_ω and $I_{2\omega}$ being the light intensities of fundamental and SH radiation, respectively) that [27]

$$\left(\frac{I_{2\omega}}{I_{\omega}^{2}}\right)_{sat}=\left(\frac{\mu_0}{\varepsilon_0}\right)^{1/2}\frac{16n_{2\omega}e^{2}\kappa}{n_{\omega}^{2}\hbar\omega^{3}m_{e}}\leq 10^{-16}m^{2}/W, \tag{5}$$

where n_ω and $n_{2\omega}$ are the refractive indexes, e and m_e are the elementary charge and mass of the electron, respectively, $\hbar$ is the reduced Planck constant, and κ is a factor related to the ionized electron momentum which is less but of the order unity.

It is interesting to note that the saturation formula similar to Eq. (5) was derived by an other purely phenomenological cumulative local-response model [29] as well.

A number of publish experiments on self-organized SHG with the self-seeding and the external SH seeding was evaluated and it has been found that the experimentally determined value of $(I_{2\omega}/I_{\omega}^{2})_{sat}$ ranges from about 0.2×10^{-16} m^2/W to 2×10^{-16} m^2/W [30].

It seems very probably that the self-saturation of self-organized SHG is the dominant mechanism behind moving the photoinduced $\chi^{(2)}$ grating backwards if the preparation process runs sufficiently long time, which was observed by Lacerda et al. [31].

Another important feature of self-organized SHG is the phase behaviour of generated SH light. The mutual phase shift of the SH radiation generated at the photoinduced $\chi^{(2)}$ grating and the SH seeding (input) radiation, $\Delta\varphi_{2\omega} = \varphi_{2\omega}^{gen} - \varphi_{2\omega}^{seed}$, was measured by Margulis et al. [32] for the first time. This measurement yielded the phase shift of about 90°, which is precisely the wrong value. However, the SH seeding intensity was relatively strong and approached the saturation value in this case. For somewhat lower SH seeding intensities other values of $\Delta\varphi_{2\omega}$ were reported as well [33-35].
In 1996 Lambelet and Feinberg [35] performed a unique experiment in which they measured both the power and the phase of SH output radiation emerging from a short germanium-doped glass fiber being conditioned to SHG in the whole course of preparation process. Using these experimental results, we have computed the evolution of $\Delta\varphi_{2\omega}(t)$ in one-hour preparation time interval [36]. Our computation has shown that $\Delta\varphi_{2\omega}(t)$ monotonously increased from its initial value $\Delta\varphi_{2\omega}(t) \approx 70°$ and approached $\Delta\varphi_{2\omega,sat}(t) \approx 90°$ in the saturation state. Evidently, the mutual phase shift $\Delta\varphi_{2\omega} = \varphi_{2\omega}^{gen} - \varphi_{2\omega}^{seed}$ cannot be considered constant but it varies in the course of SHG preparation process.
Taking into account both the saturation and the phase behaviour of self-organized SHG we have proposed a new phenomenological model that describes the self-organized $\chi^{(2)}$ grating formation at the interaction of fundamental and SH waves by means of $\chi^{(2)}$ encoding function as follows [26],

$$\chi^{(2)}(z,t) = -\exp(i\Delta kz)\int_0^t F\left[I_\omega(z,t-\tau), I_{2\omega}(z,t-\tau),\tau\right] A_\omega^2(z,t-\tau)\, A_{2\omega}^*(z,t-\tau)\, d\tau + c.c., \quad (6)$$

where the $\chi^{(2)}$ encoding function $F(z,t-\tau)$ is complex, it is dependent on the instantaneous light intensities I_ω and $I_{2\omega}$ and it varies in the course of SHG preparation process, as well as along the fiber.
In the approximation of nondepleted pump, $A_\omega = \text{const.}$, the spatial evolution of SH field amplitude can be described by means of simple first-order differential equation

$$\frac{\partial A_{2\omega}(z,t)}{\partial z} = -i\left(\frac{\mu_0}{\varepsilon_0}\right)^{1/2} \frac{\omega}{n_{2\omega}} \chi^{(2)}(z,t)\, A_\omega^2(z,t). \quad (7)$$

The solution of Eqs. (6) and (7) together with the discussion was published in [26]. It has been shown that only the positive imaginary part of $\chi^{(2)}$ encoding function, $\mathrm{Im}[F(z,t-\tau)]$, contributes effectively to the self-organized $\chi^{(2)}$ grating formation in the course of self-organized SHG [26]. There is a general tendency of self-organized $\chi^{(2)}$ grating formation to saturate, meaning that $F(z,t-\tau)$ rotates in the complex plane towards the real axis in the course of $\chi^{(2)}$ grating formation and, simultaneously, its absolute value decreases and approaches zero in the saturation state [26].
An interesting question occurs what happens if the SH seeding radiation exceeds the saturation value $I_{2\omega}^{seed}/I_\omega^2 > \left(I_{2\omega}/I_\omega^2\right)_{sat}$. In Anderson's et al. picture [27], an unstable

state is created and the interaction is expected to have a tendency to saturate. This can happen due to the energy transfer from the SH radiation into the fundamental radiation, namely, the parametric down conversion (PDC). The $\chi^{(2)}$ encoding function $\mathsf{F}(z,t-\tau)$ possesses a negative imaginary part in this case, it rotates towards the real axis and, simultaneously, its absolute value decreases during the $\chi^{(2)}$ preparation process.

It seems very likely that the $\chi^{(2)}$ gratings were already formed owing to the self-organized PDC in some experiments reported [25, 37, 38], if the SH seeding radiation exceeded the saturation value.

The first aimed experiment on self-organized PDC was performed by P. Chmela and O. Samek in the Laboratory of Nonlinear Optics, Physics Department of Pontifícia Universidade Católica in Rio de Janeiro.

The experiment was carried out with the Nd:YAG laser operated Q-switched (at 1.2 kHz repetition rate) and mode-locked (at 100 MHz). The infrared radiation at 1.064 μm was frequency doubled in KTP crystal and both the infrared radiation and the SH green radiation at 0.532 μm were launched into a germanium-doped 25-cm long standard telecommunication optical fiber. The Average output power of fundamental infrared radiation propagating in the fiber was measured to be about $P_{\omega}^{output}(0) \approx 17\mathrm{mW}$ and the SH power propagating in the fiber was about $P_{2\omega}^{output}(0) \approx 8\mathrm{mW}$ at the beginning of preparation process.

The relative output powers of fundamental and SH radiations measured in 5 minutes intervals are introduced in Table 1.

TABLE 1. Relative output powers of fundamental and SH output radiations in the course of $\chi^{(2)}$ grating formation owing to the self-organized parametric down conversion.

Preparation time [min]	$P_{\omega}^{output}(t)/P_{\omega}^{output}(0)$	$P_{2\omega}^{output}(t)/P_{2\omega}^{output}(0)$
0	1	1
5	1.05	0.84
10	1.1	0.78
15	1.1	0.8
20	1.1	0.73
25	1.14	0.7
30	1.14	0.69
35	1.15	0.68
40	1.15	0.68

It is evident from Table 1 that the SH output power decreased whilst the fundamental output power increased in the course of self-organized second-order nonlinear optical interaction, which clearly demonstrates that the self-organized PDC took place in the conditioned fiber. Consequently, the SH input light intensity was above the saturation value in our experiment. Unfortunately, not all parameters were known to calculate the light intensities of interacting radiations in the fiber.

The prepared fiber was tested by launching the infrared radiation alone. The slight green radiation was generated in the prepared fiber which was considerably weaker than we

had expected. Maybe the relatively effective PDC in the course of fiber preparation process happened owing to the instantaneous self-shaped $\chi^{(2)}$ grating which mostly disappeared after switching off the fundamental and SH radiation fields and only a residual permanent $\chi^{(2)}$ grating survived. Further experiments must be performed to elucidate this problem.
Nevertheless, the self-organized PDC was unambiguously experimentally demonstrated for the first time.
We are of the opinion that a considerable increasing of the typical ratio $I_{2\omega}/I_\omega^2$ above the saturation value, $\left(I_{2\omega}/I_\omega^2\right)_{seed} >> 10^{-16} m^2/W$, will provide a very effective self-organized PDC that could radically reverse the further development of the research on photoinduced self-organized nonlinear optical phenomena in doped-glass waveguide structures. Many potential applications are envisioned, especially for new kinds of self-shaped amplifiers or frequency down converters.

References

[1] U. Österberg and W. Margulis, Optics Lett. 11 (1986), 516; 12 (1987), 57.
[2] R. H. Stolen, Second Harmonic Generation in Optical Fibers, in: *Nonlinear Waves in Solid State Physics, NATO ASI Series* (Eds. A. D. Boardman, T. Twardowski, and M. Bertollotti), Plenum Press, New York 1990, pp. 297 – 324.
[3] International Workshop on Photoinduced Self-Organized Effects in Optical Fiber (Ed. F. Ouellette), *Proc. SPIE* 1516 (1991).
[4] Photosensitivity and Self-Organization in Optical Fibers and Waveguides (Ed. F. Ouellette), *Proc. SPIE* 2044 (1993).
[5] P. Chmela, Self-Organized Nonlinear Optical Phenomena in Optical Fibers, in: *Modern Nonlinear Optics, Part 1* (Eds. M. Evans and S. Kielich), Wiley, New York 1993, pp. 249 – 305.
[6] E. M. Dianov and D. S. Starodubov, *Optical Fiber Technology 1* (1994), 3.
[7] F. P. Payne, *Electron. Lett. 23* (1987), 1215.
[8] P. Chmela, *Optics Lett. 13* (1988), 669.
[9] V. Mizrahi, Y. Hibino, and G. I. Stegeman, *Optics Commun. 78* (1990), 283.
[10] Kamal, D. A. Weinberger, and W. H. Weber, *Optics Lett.* 15 (1990), 613.
[11] Kamal, M. L. Stock, A. Szpak, C. H. Thomas, D. A. Weinberger, M. Frankel, J. Nees, K. Ozaki, and J. A. Waldmanis, *Technical Digest of OSA Annual Meeting, Optical Society of America*, Washington D.C. 1990, paper PD 25.
[12] P. A. Franken and J. F. Ward, *Rev. Modern Physics* 35 (1963), 23.
[13] S. Kielich, *IEEE J. Quant. Electronics* QE-5 (1969), 562.
[14] J. A. Armstrong, N. Bloembergen, J. Ducuing, and P. S. Pershan, *Phys. Rev.* 127 (1962), 1918.
[15] M. O. Okada, K. Kakizawa, and S. Ieiri, *Optics Commun.* 18 (1976), 331; Japanese J. Appl. Phys. 16 (1977), 55.
[16] J. Petráček and P. Chmela, *Optical and Quantum Electronics* 26 (1994), 609.
[17] R. H. Stolen and H. W. K. Tom, *Optics Lett.* 12 (1987), 585.
[18] Lesche, *J. Opt. Soc. Am. B* 7 (1990), 53.
[19] N. M. Lawandy, *Phys. Rev. Lett.* 65 (1990), 1745.
[20] N. M. Lawandy and R. L. MacDonald, *J. Opt. Soc. Am. B* 8 (1991), 1307.
[21] T. E. Tsai and D. L. Griscom, *SPIE Proc.* 1516 (1992), 14.
[22] E. M. Dianov, P. G. Kazanski, and D. Yu. Stepanov, *Sov. J. Quantum Electron.* 19 (1989), 575.
[23] E. M. Dianov, P. G. Kazanski, and D. Yu. Stepanov, *Kvant. Electron.* 17 (1990), 926.
[24] E. M. Dianov, P. G. Kazanski, D. S. Starodubov, and D. Yu. Stepanov, *Sov. Lightwave Commun.* 2 (1992), 83.
[25] M. Krol, D. J. Giovani, W. Pleibel, and R. H. Stolen, *Optics Lett.* 18 (1993), 1220.
[26] P. Chmela and J. Petráček, *Optics Commun.* 156 (1998), 374.

[27] Z. Anderson, V. Mizrahi, and J. E. Sipe, *Optics Lett.* 16 (1991), 796.
[28] N. M. Lawandy, T. J. Driscol, and C. L. Adler, *IEEE Proceedings J.* 139 (1992), 133.
[29] P. Chmela, J. Petráček, A. Romolini, T. Pascucci, and R. Falciai, *Optics Commun.* 113 (1994), 305.
[30] P. Chmela, J. Petráček, A. Romolini, T. Pascucci, and R. Falciai, *Optical Fiber Technology* 1 (1995), 352.
[31] M. M. Lacerda, I. C. S. Carvalho, W. Margulis, and B. Lesche, *Electron. Lett.* 30 (1994), 732.
[32] W. Margulis, I. C. S. Carvalho, and J. P. von der Weid, *Optics Lett.* 14 (1989), 700.
[33] M. Dianov, P. G. Kazansky, D. S. Starodubov, and D. Yu. Stepanov, *Sov. Lightwave Commun.* 1 (1991), 395.
[34] V. Dominic and J. Feinberg, *SPIE Proc.* 2044 (1993), 223.
[35] P. Lambelet and J. Feinberg, *Optics Lett.* 21 (1996), 925.
[36] P. Chmela, J. Petráček, and K. Gniadek, *Optical and Quantum Electronics* 30 (1998), 271.
[37] Demouchy and G. R. Boyer, *Optics Commun.* 101 (1993), 385.
[38] Demouchy, *Optics Commun.* 101 (1993), 515.

DYNAMICS OF THE ELECTRO-OPTIC PROPERTIES OF DYE-CONTAINING POLYMERS

P.M. JOHANSEN[1], K.G. JESPERSEN[1], T.G. PEDERSEN[2]

1) *Optics and Fluid Dynamics Department, Risø National Laboratory, DK-4000 Roskilde, Denmark, Tel. +45 46774511, fax +45 46774588,* per.michael.johansen@risoe.dk

2) *Institute of physics, Aalborg University, Pontoppidanstræde 103, DK-9220 Aalborg Øst, Denmark, Tel. +45 96359228, fax +45 98156502,* tgp@physics.auc.dk

Abstract

The electro-optic response of polymer systems arises mainly from reorientation of the nonlinear optical chromophores that interact strongly with an applied electric field due to their large dipole moment (typically 4-8 debyes). Experimentally, the electro-optics coefficients of a polymer can be determined using a standard Teng-Man setup in which the polymer is sandwiched between electrodes in a reflection geometry. By making one electrode transparent the change in optical properties can be directly monitored. Theoretically, the effect can be explained using a statistical model based on a Smoluchowsky equation (rotation diffusion model). In this talk we shall discuss, theoretically and experimentally, the response to both dc and ac applied electric fields. Also, the effects from dipole-dipole correlations and the effects from the viscous polymer matrix will be covered.

Full article not available

A. Driessen (ed.), Nonlinear Optics for the Information Society, 193.

Author index

MIX
Papier aus verantwortungsvollen Quellen
Paper from responsible sources
FSC® C105338

If you have any concerns about our products,
you can contact us on
ProductSafety@springernature.com

In case Publisher is established outside the EU,
the EU authorized representative is:
Springer Nature Customer Service Center GmbH
Europaplatz 3, 69115 Heidelberg, Germany

Printed by Libri Plureos GmbH
in Hamburg, Germany